Acoustics for Engineers

Troy Lectures

Jens Blauert and Ning Xiang

Acoustics for Engineers

Troy Lectures

Second Edition

 Springer

Prof. Jens Blauert, Dr.-Ing., Dr. Tech. h.c.
Institute of Communication Acoustics
Ruhr-University Bochum
44780 Bochum
Germany
E-mail: jens.blauert@rub.de

Prof. Ning Xiang, Ph.D.
School of Architecture
Rensselaer Polytechnic Institute
110, 8th Street, Troy, NY 12180-3590
USA
E-mail: xiangn@rpi.edu

ISBN 978-3-642-42477-9   ISBN 978-3-642-03393-3 (eBook)

DOI 10.1007/978-3-642-03393-3

Typesetting: Data supplied by the authors

Production: Scientific Publishing Services Pvt. Ltd., Chennai, India

Cover Design: eStudio Calamar, Steinen-Broo

Printed in acid-free paper

9 8 7 6 5 4 3 2 1

springer.com

Preface

This book provides the material for an introductory course in engineering acoustics for students with basic knowledge in mathematics. It is based on extensive teaching experience at the university level.

Under the guidance of an academic teacher it is sufficient as the sole textbook for the subject. Each chapter deals with a well defined topic and represents the material for a two-hour lecture. The chapters alternate between more theoretical and more application-oriented concepts.

For the purpose of self-study, the reader is advised to use this text in parallel with further introductory material. Some suggestions to this end are given in Appendix 15.3.

The authors thank *Dorea Ruggles* for providing substantial stylistic refinements. Further thanks go to various colleagues and graduate students who most willingly helped with corrections and proof reading and, last but not least, to the reviewers of the 1st edition, particularly to Profs. *Gerhard Sessler* and *Dominique J. Chéenne*. Nevertheless, the authors assume full responsibility for all contents.

For the 2nd edition, typos have been corrected and a number of figures, notations and equations have been edited to increase the clarity of presentation. Further, a collection of problems has been included. Solutions to the problems will be provided on a peer-to-peer basis via the internet – see Appendix 15.4 for the link.

Bochum and Troy,   *Jens Blauert*
June 2009   *Ning Xiang*

Contents

1

Introduction

Human beings are usually considered to predominantly perceive their environment through the visual sense – in other words, humans are conceived as *visual beings*. However, this is certainly not true for inter-individual communication.

In fact, it is audition and not vision that is the most relevant social sense of human beings. The auditory system is their most important communication organ. Please take as proof that blind people can be educated much more easily than deaf ones. Also, when watching TV, an interruption of the sound is much more distracting than an interruption of the picture. Particular attributes of audition compared to vision are the following.

- In audition, communication is compulsory. The ears cannot be closed by reflex like the eyes
- The field of hearing extends to regions all around the listener – in contrast to the visual field. Further, it is possible to listen behind optical barriers and in darkness

These special features, among other things, lead many engineers and physicists, particularly those in the field of communication technology, to a special interest in acoustics. A further reason for the affinity of engineers and physicists to acoustics is based on the fact that many physical and mathematical foundations of acoustics are usually well known to them, such as mechanics, electrodynamics, vibration, waves, and fields.

1.1 Definition of Three Basic Terms

When you work your way into acoustics, you will usually start with the phenomenon of hearing. Actually, the term *acoustics* is derived from the Greek verb ακούειν [akúɪn], which means *to hear*. We thus start with the following definition.

> *Auditory event* ... An auditory event is something that exists as heard. It becomes actual in the act of hearing. Frequently used synonyms are *auditory object, auditory percept,* and *auditory sensation*

Consequently, the question arises of when do auditory events appear? As a rule, we hear something when our auditory system interacts via the ears with a medium that moves mechanically in the form of vibrations and/or waves. Such a medium may be a fluid like air or water, or a solid like steel or wood. Obviously, the phenomenon of hearing usually requires the presence of mechanic vibration and/or waves. The following definition follows this line of reasoning.

> *Sound* ... Sound is mechanic vibration and/or mechanic waves in elastic media

According to this definition, sound is a purely physical phenomenon. Please be warned, however, that the term *sound* is also sometimes used for auditory events, particularly in sound engineering and sound design. Such an ambiguous usage of the term is avoided in this book.

It should be briefly mentioned that vibrations and waves can often be mathematically described by differential equations – see Chapter 2. Vibration requires a common differential equation since the dependent variable is a function of time, while waves require partial ones since the dependent variable is a function of both time and space. Further, it should be noted that, although rare, auditory events may happen without sound being present, as with tinnitus. In turn, there may be no auditory events in the presence of sound, for example, for deaf people or when the frequency range of the sound is not in the range of hearing. Sounds can be categorized in terms of their frequency ranges – listed in Table 1.1 .

Table 1.1. Sound categories by frequency range

Sound category	Frequency range
Audible sound	$\approx 16\,\text{Hz}$–$16\,\text{kHz}$
Ultrasound	$> 16\,\text{kHz}$
Infrasound	$< 16\,\text{Hz}$
Hypersound	$> 1\,\text{GHz}$

The interrelation of auditory events and sound is captured by the following definition of acoustics.

Acoustics ... Acoustics is the science of sound and of its accompanying auditory events

This book deals with *engineering acoustics*. Synonyms for engineering acoustics are *applied acoustics* and *technical acoustics*.

1.2 Specialized Areas within Acoustics

In Fig. 1.1 (a) we present a schematic of a transmission system as it is often used in communication technology. A source renders information that is fed into a sender in coded form and transmitted over a channel. At the receiving end, a receiver picks up the transmitted signals, decodes them, and delivers the information to its final destination, the information sink.

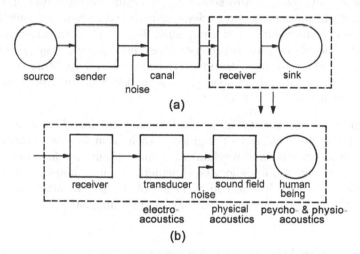

Fig. 1.1. Schematic of a transmission system **(a)** general, **(b)** electroacoustic transmission system – receiving end

In Fig. 1.1 (b) the schematic has been modified so as to describe the receiving end of a transmission chain with acoustics involved. This schematic can help distinguish between major areas within engineering acoustics. The transmission channel delivers signals that are essentially chunks of electric energy. These signals are picked up by the receiver and fed into an energy transducer that transforms the electric energy into mechanic (acoustic) energy. The acoustic signals are then sent out into a sound field where they propagate to the listener. The listener receives them, decodes them and processes the information. Please also note that, in addition to the desired signals, undesired noise may enter the system at different points.

The main areas of acoustics are as follows. The field that deals with the transduction of acoustic energy into electric energy, and vice versa, is called *electroacoustics*. The field that deals with the radiation, propagation, and reception of acoustic energy is called *physical acoustics*. The fields that deal with sound reception and auditory information processing by human listeners are called *psychoacoustics* and *physiological acoustics*. The first of these focuses on the relationship between the sound and the auditory events associated with it, and the second deals with sound-induced physiological processes in the auditory system and brain.

Acoustics as a discipline is usually further differentiated due to practical considerations. The cover labels of the sessions at a recent major acoustics conference are illustrative of the broadness of the field:

> Active acoustic systems, audiological acoustics, audio technology, building acoustics, bioacoustics, electroacoustics, vehicle acoustics, evaluation of noise, hydro-acoustics, structure-borne sound, noise propagation, noise protection, effects of noise, education in acoustics, acoustic-measurement engineering, musical acoustics, medical acoustics, numerical acoustics, physical acoustics, psychoacoustics, room acoustics, virtual reality, vibration technology, acoustic and auditory signal processing, speech-and-language processing, flow acoustics, ultrasound, virtual acoustics

Accordingly, a large variety of professions can be found that deal with acoustics, including a variety of engineers, such as audio, biomedical, civil, electrical, environmental and mechanical engineers. Further, for example, administrators, architects, audiologists, designers, ear-nose-and-throat-doctors, lawyers, managers, musicians, computer scientists, patent attorneys, physicists, physiologists, psychologists, sociologists and linguists.

1.3 About the History of Acoustics

Acoustics is a very old science. *Pythagoras* already knew, around 500 BC, of the quantitative relationship between the length of a string and the pitch of its accompanying auditory event. In 1643, *Torricelli* demonstrated the vacuum experimentally and showed that there is no sound propagation in it. At the end of the 19th century, classical physical acoustics had matured. The book "The Theory of Sound" by *Rayleigh* 1896, is considered to be an important reference even today.

At about the same time, basic inventions in acoustical communication technology were made, including the telephone (*Reis* 1867), television (*Nipkow* 1884) and tape recording (*Ruhmer* 1901). It was only after the independent invention of the vacuum triode by *von Lieben* and *de Forest* in 1910, which made amplification of weak currents possible, that modern acoustics enjoyed a real up-swing through applications such as radio broadcast since 1920, sound-on-film since 1928, and public-address systems with loudspeakers since 1924.

Starting in about 1965, computers made their way into acoustics, making effective signal processing and interpretation possible and leading to advanced applications such as acoustical tomography, speech-and-language technology, surround sound, binaural technology, auditory displays, mobile phones, and many others. Acoustics in the context of the information and communication technologies and sciences is nowadays called *communication acoustics*.

In this book we shall, however, concentrate on the classical aspects of *engineering acoustics*, particularly on *physical acoustics* and *electroacoustics*. To this end, we shall make use of the following theoretical tools: the theory of electric and magnetic processes, the theory of signals, vibrations and systems, and the theory of waves and fields.

1.4 Relevant Quantities in Acoustics

The following quantities are of particular relevance in acoustics.

- *Displacement, elongation*

  $\vec{\xi}$, in [m] ... displacement of an oscillating particle from its resting position

- *Particle velocity*

  $\vec{v}$, in [m/s] ... alternating velocity of an oscillating particle

- *Sound pressure*

  p, in [N/m^2 = Pa] ... alternating pressure as caused by particle oscillation[1]

- *Sound intensity*

  $\vec{I}$, in [W/m^2] ... sound power per effective area, $A_\perp$, that is the area component perpendicular to the direction of energy propagation

- *Speed of sound*

  $\vec{c}$, in [m/s] ... propagation speed of a sound wave[2]

The superscribed arrows denote vectors, but we shall use them only when the vector quality is of relevance. Otherwise we use the magnitude, $c = |\vec{c}|$.

Since sound is essentially vibrations and waves, the quantities ξ, v, and p are periodically alternating quantities. According to *Fourier*, they can be

---

[1] $1\,\mathrm{Pa} = 1\,\mathrm{N/m}^2 = 1\,\mathrm{kg/(ms}^2) = 1\,(\mathrm{Ws})/\mathrm{m}^3$
[2] Warning: $\vec{c}$ must not be mistaken as a particle velocity!

decomposed into sinusoidal components. These components can then be described in complex notation – see Appendix 15.3. Quantities known to be complex are underlined in this book.

In acoustics, there are the following three definitions of impedances.

- *Field impedance*

$$\underline{Z}_f = \frac{\underline{p}}{\underline{v}}, \quad \text{in } [\frac{\text{Ns}}{\text{m}^3}] \tag{1.1}$$

- *Mechanic impedance*

$$\underline{Z}_{\text{mech}} = \frac{\underline{F}}{\underline{v}}, \quad \text{in } [\frac{\text{Ns}}{\text{m}}] \tag{1.2}$$

- *Acoustic impedance*

$$\underline{Z}_a = \frac{\underline{p}}{\underline{v}\,A}, \quad \text{in } [\frac{\text{Ns}}{\text{m}^5}] \tag{1.3}$$

with $\underline{F}$ being the force, $\underline{F} = \underline{p}\,A$, and $\underline{q}$ being the so-called volume velocity, $\underline{q} = \underline{v}\,A$. The different kinds of impedances can be converted into each other, provided that the effective radiation area of the sound source, A, is known.

Please note that impedances represent the complex ratio of two quantities, the product of which forms a power-related quantity.

1.5 Some Numerical Examples

In order to derive some illustrative numerical examples, we consider a plane wave in air[3]. A plane wave is a wave where all quantities are invariant across areas perpendicular to the direction of wave propagation. The field impedance in a plane wave is a quantity that is specific to the medium and is called the *characteristic field impedance*, $\underline{Z}_w$ – see Section 7.4. Disregarding dissipation, this is a real quantity. In air we have $Z_{w,\text{air}} \approx 412\,\text{Ns/m}^3$ under standard conditions.

- *Sound pressure*

    Sound pressure at the threshold of discomfort (maximum sound pressure),

$$p_{\text{max, rms}} \approx 10^2\,\text{N/m}^2 = 100\,\text{Pa} \tag{1.4}$$

---

[3] We present rms-values rather than peak values in the following synopsis to account for all kinds of finite-power sounds, such as noise, speech and music, besides sinusoidal sounds. See Appendix 15.4 for the definition of rms. Peak values of sinusoidal signals – as usually used in complex notations throughout this book – exceed their rms-values by a factor of $\sqrt{2}$, that is $\hat{x} = \sqrt{2}\,x_{\text{rms}}$

Sound pressure at a normal conversation level at $1\,\text{m}$ distance from the talker (normal sound pressure),

$$p_{\text{normal, rms}} \approx 0.1\,\text{N/m}^2 = 100\,\text{mPa} \tag{1.5}$$

Sound pressure at $1\,\text{kHz}$ at the threshold of hearing (minimum sound pressure),

$$p_{\text{min, rms}} \approx 2 \cdot 10^{-5}\,\text{N/m}^2 = 20\,\mu\text{Pa} \tag{1.6}$$

For reference: The static atmospheric pressure under normal conditions is about $10^5\,\text{N/m}^2 = 1000\,\text{hPA} \,\hat{=}\, 1\,\text{bar}$

- *Particle velocity*

  The following particle velocities appear with the above sound pressures, considering the relationship $\underline{p} = Z_{\text{w, air}}\,\underline{v}$.

  Maximum particle velocity, $v_{\text{max, rms}} \approx 0.25\,\text{m/s}$

  Normal particle velocity, $v_{\text{normal, rms}} \approx 25 \cdot 10^{-5}\,\text{m/s}$

  Minimum particle velocity, $v_{\text{min, rms}} \approx 5 \cdot 10^{-8}\,\text{m/s}$

  For reference: The speed of sound in air is $c \approx 340\,\text{m/s}$

- *Particle displacement*

  The relationship between particle velocity and particle displacement is frequency dependent as follows, $\xi(t) = \int v(t)\,dt$, or, in complex notation, $\underline{\xi} = \underline{v}/j\omega$. A comparison thus requires selection of a specific frequency. We have chosen $1\,\text{kHz}$ here. With this presupposition we get,

  Maximum particle displacement, $\xi_{\text{max, rms}} \approx 4 \cdot 10^{-5}\,\text{m}$

  Normal particle displacement, $\xi_{\text{normal, rms}} \approx 4 \cdot 10^{-8}\,\text{m}$

  Minimum particle displacement, $\xi_{\text{min, rms}} \approx 8 \cdot 10^{-12}\,\text{m}$

  For reference: The diameter of a hydrogen atom is $10^{-10}\,\text{m}$. Actually, for the small displacements near the threshold of hearing it becomes questionable whether consideration of the medium as a continuum is still valid

It is also worth noting here that the particle displacements due to the *Brownian* molecular motion are only one order of magnitude smaller than those induced by sound at the threshold of hearing. Thus the auditory system works definitely at the brink of what makes sense physically. If the system where only a little more sensitive, one could indeed "hear the grass growing."

1.6 Levels and Logarithmic Frequency Intervals

As shown above, the range of sound pressures that must be handled in acoustics is at least $1:10{,}000{,}000$, which is $1:10^7$. This leads to unhandy numbers when describing sound pressures and sound-pressure ratios. For this and other reasons, a logarithmic measure called the *level* is frequently used. The other reasons for its use are the following.

- Equal relative modifications of the strength of a physical stimulus lead to equal absolute changes in the salience of the sensory events, which is called the *Weber-Fechner* law and can be approximated by a logarithmic characteristic

- When connecting two-port elements in chain (cascade), the overall level reduction (attenuation) between input and output turns out to be the sum of the attenuations of each element

The following level definitions are common in acoustics, with $\lg = \log_{10}$.

- *Sound-intensity level*

$$L_\mathrm{I} = 10 \lg \frac{|\vec{I}|}{I_0}\,\mathrm{dB}\,,\ \ \text{with } I_0 = 10^{-12}\,\mathrm{W/m^2} \text{ as reference} \qquad (1.7)$$

- *Sound-pressure level*

$$L_\mathrm{p} = 20 \lg \frac{p_\mathrm{rms}}{p_{0,\mathrm{rms}}}\,\mathrm{dB}\,,\ \ \text{with } p_{0,\mathrm{rms}} = 2 \cdot 10^{-5}\,\mathrm{N/m^2} = 20\,\mu\mathrm{Pa}$$
$$\text{as reference} \qquad (1.8)$$

- *Sound-power level*

$$L_\mathrm{P} = 10 \lg \frac{|P|}{P_0}\,\mathrm{dB}\,,\ \ \text{with } P_0 = 10^{-12}\,\mathrm{W} \text{ as reference} \qquad (1.9)$$

The reference levels are internationally standardized, and the first two roughly represent the threshold of hearing at $1\,\mathrm{kHz}$. Other references may be used, but in such cases the respective reference must be noted, for example, in the form $L = 15\,\mathrm{dB}$ re $100\,\mu\mathrm{Pa}$. The symbol used to signify levels computed with the above definitions is [dB], which stands for deciBel, named after *Alexander Graham Bell*. Another unit-like symbol based on the natural logarithm, $\log_\mathrm{e} = \ln$, the Neper [Np], is also used to express level, particularly in transmission theory. Levels in Neper can be converted into levels in deciBel as follows, $L\,[\mathrm{Np}] = 8.69\,L\,[\mathrm{dB}]$[4]

---

[4] Note that deciBel [dB] and Neper [Np] are no units in the strict sense but letter symbols indicating a computational process. When used in equations, their dimension is one

In the case of intensity and power levels, it should be noted that the levels describe ratios of the magnitudes of intensity and/or power. These magnitudes read as follows in complex notation, taking the intensity as example – see Appendix 15.2.

$$|\underline{\vec{I}}| = \left| I\, e^{j\omega\,(\phi_\mathrm{p}-\phi_\mathrm{q})} \right| = \frac{1}{2}\left| \underline{p}\,\underline{q}^* \right|. \tag{1.10}$$

For practical purposes, it is useful to learn some level differences by heart. A few important examples are listed in the Table 1.2. By knowing these values, it is easy to estimate level differences. For instance, the sound-pressure ratio of $1:2000 = (1:1000)\,(1:2)$ corresponds to -60 dB - 6 dB = -66 dB.

Table 1.2. Some useful level differences

Ratio of sound pressure	Ratio of sound intensity or power
$\sqrt{2}:1 \approx 3\,\mathrm{dB}$	$\sqrt{2}:1 \approx 1.5\,\mathrm{dB}$
$2:1 \approx 6\,\mathrm{dB}$	$2:1 \approx 3\,\mathrm{dB}$
$3:1 \approx 10\,\mathrm{dB}$	$3:1 \approx 5\,\mathrm{dB}$
$5:1 \approx 14\,\mathrm{dB}$	$5:1 \approx 7\,\mathrm{dB}$
$10:1 = 20\,\mathrm{dB}$	$10:1 = 10\,\mathrm{dB}$

In order to compute the levels that add up when more than one sound source is active, one has to distinguish between (a) sounds that are coherent, such as stemming from loudspeakers with the same input signals, and (b) those that are incoherent, such as originating from independent noise sources like vacuum cleaners. Coherent sounds interfere but incoherent ones do not. Consequently, we end up with the following two formulas for summation.

• *Addition of coherent (sinusoidal) sounds*

$$L_\Sigma = 20\lg\left(\frac{\frac{1}{\sqrt{2}}\,|\underline{p}_1 + \underline{p}_2 + \underline{p}_3 + \cdots + \underline{p}_\mathrm{n}|}{p_{0,\mathrm{rms}}} \right)\ \mathrm{dB} \tag{1.11}$$

• *Addition of incoherent sounds*

$$L_\Sigma = 10\lg\left(\frac{|\underline{\vec{I}}_1| + |\underline{\vec{I}}_2| \cdots + |\underline{\vec{I}}_\mathrm{n}|}{I_0} \right)\ \mathrm{dB} \tag{1.12}$$

We see that inter-signal phase differences need not be considered when the signals do not interfere.

• *Logarithmic frequency intervals*

What holds for the magnitude of sound quantities, namely, that their range is huge, also holds for the frequency range of the signal components. The audible frequency range is roughly considered to extend from about 16 Hz to 16 kHz in young people, which is a range of $1 : 10^3$. With high-intensity sounds, some kind of hearing may even be experienced above 16 kHz. Sensitivity to high frequencies decreases with age.

We find a logarithmic relationship also with regard to frequency. The equal ratios between the fundamental frequencies of musical sounds lead to equal musical intervals of pitch.

Therefore, a logarithmic ratio of frequencies called *logarithmic frequency interval*, Ψ, has been introduced. It is based on the logarithmus dualis, $\mathrm{ld} = \log_2$, and is of dimension one. The following four definitions are in use,

$$
\begin{aligned}
\Psi_{\mathrm{oct}} &= \mathrm{ld}\,(f_1/f_2), & &\text{in [oct] ... octave} \\
\Psi_{1/3\mathrm{rd\,oct}} &= 3\,\mathrm{ld}\,(f_1/f_2), & &\text{in } [\tfrac{1}{3}\,\mathrm{oct}] \\
\Psi_{\mathrm{semitone}} &= 12\,\mathrm{ld}\,(f_1/f_2), & &\text{in [semitone]} \\
\Psi_{\mathrm{cent}} &= 1200\,\mathrm{ld}\,(f_1/f_2), & &\text{in [cent]}
\end{aligned}
$$

These four logarithmic frequency intervals have the following relationship to each other, $1\,\mathrm{oct} = 3\,(\tfrac{1}{3}\,\mathrm{oct}) = 12\,\mathrm{semitone} = 1200\,\mathrm{cent}$. In communication engineering, decades (10 : 1) are sometimes preferred to octaves (2 : 1). Conversion is as follows: $1\,\mathrm{oct} \approx 0.3\,\mathrm{dec}$ or $1\,\mathrm{dec} \approx 3.3\,\mathrm{oct}$.

Wavelength, λ, and frequency, f, of an acoustic wave are linked by the relationship $c = \lambda\,f$. In air we have $c \approx 340\,\mathrm{m/s}$. In Table 1.3, a series of frequencies is presented with their corresponding wavelengths in air. The series is taken from a standardized octave series that is recommended for use in engineering acoustics.

Table 1.3. Wavelengths in air vs. octave-center frequencies

Octave-center frequency [Hz]	16	32	63	125	250	500	1k	2k	4k	8k	16k
Wave length in air [m]	20	10	5	2.5	1.25	0.63	0.32	0.16	0.08	0.04	0.02

It becomes clear that just in the audible range the wavelengths extend from a few centimeters to many meters. Because radiation, propagation, and reception of waves is characterized by the linear dimension of reflecting surfaces relative to the wavelength of the waves, a broad variety of different effects, including reflection, scattering and diffraction, are experienced in acoustics.

1.7 Double-Logarithmic Plots

By plotting levels over logarithmic frequency intervals, we obtain a double-logarithmic graphic representation of the original quantities. This way of plotting has some advantages over linear representations and is quite popular in acoustics[5]. Figure 1.2 (a) presents an example of a linear representation, and Fig. 1.2 (b) shows its corresponding double-logarithmic plot.

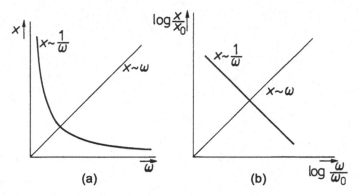

Fig. 1.2. Different representations of frequency functions. **(a)** linear, **(b)** double logarithmic

In double-logarithmic plots, all functions that are proportional to ω^y appear as straight lines since

$$x = a\,\omega^y \;\rightarrow\; \log x = \log a + y \log \omega\,. \tag{1.13}$$

For integer potencies, $y = \pm n$ with $n = 1, 2, 3, \cdots$, we arrive at slopes of $\pm n \cdot 6\,\mathrm{dB/oct}$ for sound pressure, displacement, and particle velocity, and of $\pm n \cdot 3\,\mathrm{dB/oct}$ for power and intensity. For decades the respective values are $\approx 20\,\mathrm{dB/dec}$ resp. $\approx 10\,\mathrm{dB/dec}$.

Functions with different potencies of ω are actually quite frequent in acoustics. They result from differential equations of different degree that are used to describe vibrations and waves. The slope of the lines in the plot helps estimate the order of the underlying oscillation processes.

---

[5] In network theory, double-logarithmic graphic representations are know as *Bode* diagrams

2

Mechanic and Acoustic Oscillations

When physical or other quantities vary in a specific way as a function of time, we say that they *oscillate*. A common, very broad definition of oscillation is as follows.

> *Oscillation* ... An oscillation is a process with attributes that are repeated regularly in time

Oscillating processes are widespread in our world, and they are responsible for all wave propagation such as sound, light or radio waves. The time functions of oscillating quantities can vary extensively because of the wide variation between sources. Oscillations can, for example, be initiated by intermittent sources like fog horns, sirens, the saw-tooth generator of an oscilloscope, or the blinking signal of a turning light.

A prominent category of oscillations is characterized by energy swinging between two complementary storages, namely, kinetic vs. potential energy or electric vs. magnetic energy. In many cases one can approximate these oscillating systems as linear and constant in time, which defines what is called a linear, time-invariant (LTI) system.

Mathematical treatment of LTI systems is particularly easy. A specific feature of these systems is that the *superposition principle* applies. Excitation of an LTI system by several individual excitation functions leads the system to respond according to the linear combination of the individual response to each excitation function.

The superposition principle can be written in mathematical terms as

$$y(t) = \sum_k b_k \, y_k(t) = \mathcal{F}\left\{\sum_k b_k \, x_k(t)\right\}, \quad \text{assuming } y_k(t) = \mathcal{F}\{x_k(t)\}. \quad (2.1)$$

The general exponential function with the complex frequency, $\underline{s} = \breve{\alpha} + \mathrm{j}\omega$,

$$A\,\mathrm{e}^{\underline{s}t} = A\,\mathrm{e}^{\breve{\alpha}t+\mathrm{j}\omega t} = A\,\mathrm{e}^{\breve{\alpha}t}(\cos\omega t + \mathrm{j}\sin\omega t),\qquad(2.2)$$

is an *eigen-function* of LTI systems. This means that an excitation by a sinusoidal function results in a response that is a sinusoidal function of the same frequency, although generally with a different phase and amplitude. This special feature of LTI systems is one of the reasons why sinusoidal functions play a prominent role in the analysis of LTI systems and linear oscillators.

Operations with LTI systems are often performed in what is called the *frequency domain*. To move from the time domain to the frequency domain, the time function of the excitation is decomposed by *Fourier* transforms into sinusoidal components. Each component is then sent through the system, and the time function of the total response determined by summing up all the individual sinusoidal responses and performing the inverse *Fourier* transforms.

In this book, we shall not deal with *Fourier* transforms in great detail, but the fact that all sounds can be decomposed into sinusoidal components and (re)composed from these, may be taken as a good argument for using sinusoidal excitation in LTI systems for our analyses.

2.1 Basic Elements of Linear, Oscillating, Mechanic Systems

Three elements are required to form a simple mechanic oscillator, and they include a *mass*, a *spring* and a fluidic *damper*(dashpot) – shown in Fig. 2.1.

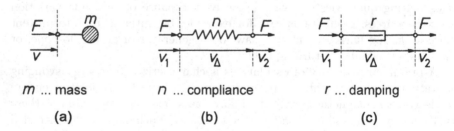

m ... mass n ... compliance r ... damping

(a)   (b)   (c)

Fig. 2.1. Basic elements of linear time-invariant mechanic oscillation systems, **(a)** mass, **(b)** spring, **(c)** fluidic damper (dashpot)

For the introduction of these elements, we make three idealizing assumptions.

(a) All relationships between the mechanic quantities displacement, ξ, particle velocity, v, force, F, and acceleration, a, are linear

(b) The characteristic features of the elements are constant
(c) We consider one-dimensional motion only

- *Mass*

  An alternating force may be applied to a solid body with mass, m – shown in Fig.2.1 (a) – so that *Newton's*[1] law holds as follows,

  $$F(t) = m\, a(t) = m\, \frac{dv}{dt} = m\, \frac{d\xi^2}{dt^2}\,. \qquad (2.3)$$

  For sinusoidal quantities we can write this law in complex notation as

  $$\underline{F} = m\, \underline{a} = j\omega\, m\, \underline{v} = -\omega^2 m\, \underline{\xi}\,. \qquad (2.4)$$

  Later we will derive that the mass stores kinetic energy. It is a *one-port* element in terms of network-theory because there is only one in/output port through which power can be transmitted. The mechanic impedance of a mass is imaginary and expressed as

  $$\underline{Z}_{\text{mech}} = j\omega\, m \qquad (2.5)$$

- *Spring*

  According to *Hook*, the following applies for linear springs[2] with a compliance of n – as seen in Fig. 2.1 (b)

  $$F(t) = \frac{1}{n}\,\xi_\Delta(t) = \frac{1}{n} \int v_\Delta(t)\, dt = \frac{1}{n} \int \left[\int a_\Delta(t)\, dt \right] dt\,. \qquad (2.6)$$

  For sinusoidal quantities in complex notation this is equivalent to

  $$\underline{F} = \frac{1}{n}\,\underline{\xi}_\Delta = \frac{1}{j\omega\, n}\,\underline{v}_\Delta = \frac{-1}{\omega^2 n}\,\underline{a}_\Delta\,. \qquad (2.7)$$

  The spring stores potential energy. It is a *two-port* element because it has both an input and an output port. The mechanic impedance of the spring is imaginary and equal to

  $$\underline{Z}_{\text{mech}} = \frac{1}{j\omega\, n} \qquad (2.8)$$

---

[1] *Newton's* law is valid in so-called inertial spatial coordinate systems. These are such in which a mass to which no force is applied moves with constant velocity along a linear trajectory. As origin of the coordinate system we usually use "ground", which is a mass taken as infinite. Gravitation forces are not considered here

[2] In acoustics, the compliance, n, is often preferred to its reciprocal, the stiffness, $k = 1/n$, as this leads to formula notations that engineers are more accustomed to – refer to Chapter 3

- *Damper* (Dashpot)

  A dashpot is a damping element based on fluid friction due to a viscous medium – see Fig. 2.1 (c). At a dashpot with damping (mechanic resistance), r, the following holds,

  $$F(t) = r\, v_\Delta(t) = r\, \frac{\mathrm{d}\xi_\Delta}{\mathrm{d}t} = r \int a_\Delta(t)\mathrm{d}t\,. \tag{2.9}$$

  In complex notation for sinusoidal quantities this is

  $$\underline{F} = r\, \underline{v}_\Delta = \mathrm{j}\omega\, r\, \underline{\xi}_\Delta = \frac{r}{\mathrm{j}\omega}\, \underline{a}_\Delta\,. \tag{2.10}$$

  The mechanic impedance of the damper is real and expressed as

  $$\underline{Z}_\text{mech} = r\,. \tag{2.11}$$

  The dashpot does not store energy. It consumes it through dissipation, which is a process of converting mechanic energy into thermodynamic energy, in other words, *heat*. The dashpot is a *two-port* element

2.2 Parallel Mechanic Oscillators

We now consider an arrangement where a mass, a spring and a dashpot are connected in parallel by idealized, that is, rigid and massless rods – see Fig. 2.2.

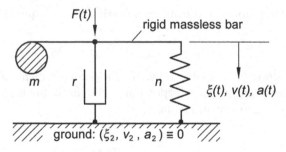

Fig. 2.2. Mechanic parallel oscillator, exited by an alternating force. The second port is grounded here for simplicity

The arrangement may be excited by an alternating force, $F(t)$, that is composed of three elements,

$$F(t) = F_\mathrm{m}(t) + F_\mathrm{n}(t) + F_\mathrm{r}(t)\,. \tag{2.12}$$

In this way, we arrive at the following differential equation,

$$F(t) = m\frac{d^2\xi}{dt^2} + r\frac{d\xi}{dt} + \frac{1}{n}\xi \quad \text{or} \quad F(t) = m\frac{dv}{dt} + r\,v(t) + \frac{1}{n}\int v(t)\,dt. \quad (2.13)$$

As only one variable, ξ or v, is sufficient to describe the state of the system, it represents what is often called a *simple oscillator*.

Please note that, for simplicity of the example, we have connected both the spring and the dashpot to ground. In this way, the quantities ξ_2 and v_2 are set to zero at the output ports, enabling the subscript Δ to be omitted.

2.3 Free Oscillations of Parallel Mechanic Oscillators

In this section we deal with the special case in which the oscillator is in a position away from its resting position, and the introduced force is set to zero, that is $F(t) = 0$ for $t > 0$. The differential equation (2.13) then converts into a homogenous differential equation as follows,

$$m\frac{d^2\xi}{dt^2} + r\frac{d\xi}{dt} + \frac{1}{n}\xi = 0. \quad (2.14)$$

The solution of this equation is called *free oscillation* or *eigen-oscillation* of the system. Trying $\underline{\xi} = e^{\underline{s}t}$, we obtain the characteristic equation[3]

$$m\,\underline{s}^2 + r\,\underline{s} + \frac{1}{n} = 0, \quad (2.15)$$

where $\underline{s}$ denotes the complex frequency. The general solution of this quadratic equation can be expressed as

$$\underline{s}_{1,2} = -\frac{r}{2m} \pm \sqrt{\frac{r^2}{4m^2} - \frac{1}{mn}} \quad \text{or} \quad \underline{s}_{1,2} = -\delta \pm \sqrt{\delta^2 - \omega_0^2}, \quad (2.16)$$

where $\delta = r/2m$ is the damping coefficient and $\omega_0 = 1/\sqrt{mn}$ the characteristic angular frequency. This general form renders the three different types of solutions, namely,

   Case (a) with $\delta < \omega_0$ ... weak damping, both roots are complex

   Case (b) with $\delta > \omega_0$ ... strong damping, both roots real, s negative

   Case (c) with $\delta = \omega_0$ ... critical damping, only one real solution of the root

---

[3] As noted in the introduction to this chapter, the general exponential function is an eigen-function of linear differential equations. It stays an exponential function when differentiated or integrated

The differential equation for a simple oscillator is of the second order, making it necessary to have two initial conditions to derive specific solutions. The following three forms of general solutions can be applied. It remains to adjust them to the particular initial conditions to finally arrive at special solutions.

- *Case* (a)

$$\underline{\xi}(t) = \underline{\xi}_1 \, e^{-\delta t} e^{-j\omega t} + \underline{\xi}_2 \, e^{-\delta t} e^{+j\omega t}, \quad \text{with } \omega = \sqrt{\omega_0^2 - \delta^2}. \quad (2.17)$$

This solution, called the *oscillating case*, describes a periodic, decaying oscillation. That we have indeed an oscillation, can best be illustrated by looking at the special case of $\underline{\xi}_1 = \underline{\xi}_2 = \underline{\xi}_{1,2}$, because there we get

$$\xi(t) = \underline{\xi}_{1,2} \, e^{-\delta t} \cos(\omega t) \quad (2.18)$$

- *Case* (b)

$$\underline{\xi}(t) = \underline{\xi}_1 \, e^{-(\delta - \sqrt{\delta^2 - \omega_0^2})t} + \underline{\xi}_2 \, e^{-(\delta + \sqrt{\delta^2 - \omega_0^2})t}. \quad (2.19)$$

This solution, called the *creeping case*, describes an aperiodic decay

- *Case* (c)

$$\underline{\xi}(t) = (\underline{\xi}_1 + \underline{\xi}_2 t) \, e^{-\delta t}. \quad (2.20)$$

This case is at the brink of both periodic and aperiodic decay. Depending on the initial conditions, it may or may not render a single swing over. It is called the *aperiodic limiting case*

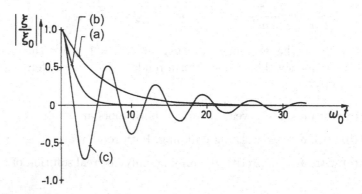

Fig. 2.3. Decays of a simple oscillator for different damping settings (schematic), **(a)** aperiodic case, **(b)** aperiodic limiting case, **(c)** oscillating case

Figure 2.3 illustrates the three cases. The fastest-possible decay below a threshold – which, by the way, is the objective when tuning the suspension of road vehicles – is achieved with a slightly subcritical damping, that is $\delta \approx 0.6\,\omega_0$.

In addition to $\delta = r/2m$, the following two quantities are often used in acoustics to characterize the amount of damping in an oscillating system,

> Q ... quality or sharpness-of-resonance factor
> T ... reverberation time

The quality factor, Q, is defined as

$$Q = \frac{\omega_0}{2\,\delta} = \frac{\omega_0\, m}{r}, \tag{2.21}$$

is a measure of the width of the peak of the resonance curve – see Section 2.4. A more illustrative interpretation is possible in the time domain when one considers that after Q oscillations a mildly damped oscillation has decreased to 4 % of its starting value, which is about what can just be visually discriminated on an oscilloscope screen.

The reverberation time, T, measures how long it takes for an oscillation to decrease by 60 dB after excitation has been stopped. At this level, velocity or displacement has decayed to one thousandth and power to one millionth of its original value. T and δ are related by $T \approx 6.9/\delta$ – refer to Section 12.5 for details.

Table 2.1 lists characteristic Q values for different kinds of technologically relevant oscillators. For comparison, in the aperiodic limiting case Q has a value of 0.5.

Table 2.1. Typical Q values for various oscillators

Type of element	Q factor
Electric oscillator of traditional construction (coil, capacitor, resistor)	$Q \approx 10^2$–10^3
Electromagnetic cavity oscillator	$Q \approx 10^3$–10^6
Mechanic oscillator, steel in vacuum	$Q \approx 5 \cdot 10^3$
Quartz oscillator in vacuum	$Q \approx 5 \cdot 10^5$
Concert hall with $T = 2\,\mathrm{s}$ at $1\,\mathrm{kHz}$	$Q \approx 900$

2.4 Forced Oscillation of Parallel Mechanic Oscillators

The exciting force was zero for free oscillations, but we will now consider the case where the oscillator is driven by an ongoing sinusoidal force,

$F(t) = \hat{F}\cos(\omega t + \phi)$, with frequency $f = \omega/2\pi$. The oscillation of the system at this point is stationary[4]. We call this mode of operation force-driven or *forced oscillation*. The mathematical description leads to an inhomogeneous differential equation as follows,

$$\hat{F}\cos(\omega t + \phi) = m\,\frac{d^2\xi}{dt^2} + r\,\frac{d\xi}{dt} + \frac{1}{n}\,\xi. \tag{2.22}$$

As we want to restrict ourselves to sinusoidal excitations here, this equation can be expressed in the following complex form

$$\underline{F} = -\omega^2 m\,\underline{\xi} + j\omega\,r\,\underline{\xi} + \frac{1}{n}\,\underline{\xi}. \tag{2.23}$$

Substitution of $\underline{\xi}$ by $\underline{v}$ yields

$$\underline{F} = j\omega\,m\,\underline{v} + r\,\underline{v} + \frac{1}{j\omega\,n}\,\underline{v}. \tag{2.24}$$

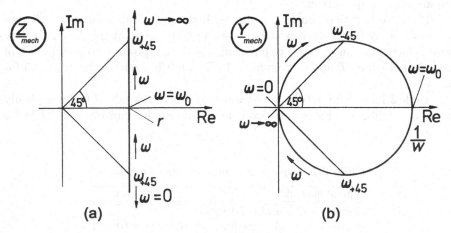

Fig. 2.4. (a) Mechanic impedance and **(b)** admittance, in the complex $\underline{Z}$ and $\underline{Y}$ planes as functions of frequency

This equation directly admits the inclusion of the mechanic impedance, $\underline{Z}_{\text{mech}}$, as well as its reciprocal, the mechanic admittance, $\underline{Y}_{\text{mech}} = 1/\underline{Z}_{\text{mech}}$, so that

$$\underline{Z}_{\text{mech}} = \frac{\underline{F}}{\underline{v}} = j\omega\,m + r + \frac{1}{j\omega\,n} \quad \text{and} \tag{2.25}$$

---

[4] We can also deal with cases where the frequency varies slowly, by assuming that a stationary state has (approximately) been reached at each instant of observation

$$\underline{Y}_{\text{mech}} = \frac{\underline{v}}{\underline{F}} = \frac{1}{j\omega\, m + r + \frac{1}{j\omega\, n}} \, . \tag{2.26}$$

Figure 2.4 illustrates the trajectories of these two quantities in the complex plane as a function of frequency. The two quantities become real at the *characteristic frequency*, ω_0. At this frequency, the phase changes signs (jumps) from positive to negative values or vice versa.

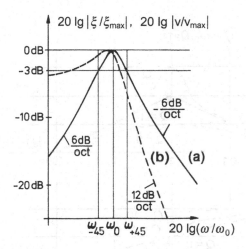

Fig. 2.5. Mechanic responses as a function of frequency for constant-amplitude forced excitation, **(a)** velocity, **(b)** elongation

When varying the frequency of excitation slowly, we observe functions of $\xi(\omega)$ and $v(\omega)$ as schematically shown in Fig. 2.5. For simple oscillators these curves have a single peak. In this example, for a case of subcritical damping with $Q \approx 2$, we have kept the exciting force constant over frequency. The course of calculations to arrive at these functions is as follows,

$$\frac{\underline{F}}{\underline{v}} = \frac{\underline{F}}{j\omega\,\underline{\xi}} = j\omega\, m + \frac{1}{j\omega\, n} + r, \tag{2.27}$$

$$\frac{\underline{\xi}}{\underline{F}} = \frac{1}{-\omega^2 m + \frac{1}{n} + j\omega\, r}, \tag{2.28}$$

$$\frac{|\underline{\xi}|}{|\underline{F}|} = \frac{1}{\sqrt{(\frac{1}{n} - \omega^2 m)^2 + (\omega r)^2}}, \tag{2.29}$$

$$\frac{|\underline{v}|}{|\underline{F}|} = \frac{1}{\sqrt{(\omega m - \frac{1}{\omega n})^2 + r^2}} \, . \tag{2.30}$$

Please note that the phase of v is decreasing and passes zero at ω_0, while the phase of ξ is also decreasing but goes through $-\pi/2$ at this point – see Fig. 2.6. Furthermore, the position of the peak for the $|v|(\omega)$ curve is exactly at the characteristic frequency, while the peak of the $|\xi|(\omega)$ curve lies slightly lower – the higher the damping, the lower the frequency at this peak! Hence, we call this peak the *resonance*. Consequently, we should properly distinguish between the terms resonance frequency and characteristic frequency.

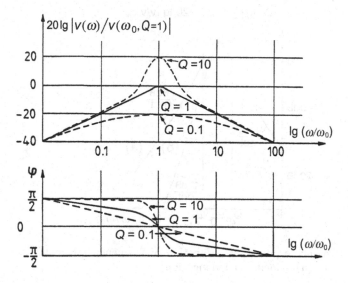

Fig. 2.6. Double-logarithmic plot of resonance curves of the particle velocity, illustrating the role of the sharpness-of-resonance factor, Q

Figure 2.6 shows the resonance curves for the particle velocity in a slightly different way to illustrate the role of the Q-factor with respect to the form of these curves. We see that the resonance peak becomes higher and more narrow with increasing Q. This is the reason that Q is termed sharpness-of-resonance factor, besides quality factor.

2.5 Energies and Dissipation Losses

To derive the energies and losses in the elements from which the oscillator is built, (2.13) is at first multiplied with $v(t)$ to arrive at what is called *instantaneous power*, namely

$$P(t) = F(t)\, v(t) = m\, \frac{dv}{dt}\, v(t) + r\, v^2(t) + \frac{1}{n}\, v(t) \int \overbrace{v(t)\, dt}^{d\xi} . \qquad (2.31)$$

Integration over time then leads to a term with the dimension energy (work) as follows,

$$W_{0,t_1} = \int_0^{t_1} F(t) \overbrace{v\,dt}^{d\xi} = m \int_0^{t_1} v\frac{dv}{dt}\,dt + r\int_0^{t_1} v^2 dt + \frac{1}{n}\int_0^{t_1} \xi v\,dt. \quad (2.32)$$

For the case that motion of the oscillator starts from resting position, that is, for $\xi_{t=0} = 0$, this can be converted into the form

$$\int_0^{\xi_1} F(t)\,d\xi = \frac{1}{2}m v_1^2 + r\int_0^{t_1} v^2 dt + \frac{1}{2}\frac{1}{n}\xi_1^2. \quad (2.33)$$

The left term denotes the energy that is fed into the system. The terms on the right side of the equality sign stand, from left to right, for the kinetic energy of the mass, the frictional losses (dissipation) in the dashpot, and the potential energy in the spring.

For discussion we start with the case of no losses, that is, when $r = 0$. In this case the total energy in the system does not change. It simply swings between the mass and spring. These relationships can be expressed as

$$W = \frac{1}{2}m v(t)^2 + \frac{1}{2n}\xi(t)^2. \quad (2.34)$$

At the instant that $\xi = 0$, all energy is potential, and when we have $v = 0$, all energy is kinetic. In mathematical terms this is

$$W(\xi = 0) = \frac{1}{2}m\hat{v}^2 = W(v = 0) \hat{=} \frac{1}{2n}\hat{\xi}^2. \quad (2.35)$$

When losses are present due to friction, that is, when $r \neq 0$, the stationary state must be preserved with a driving force. Recall that we discuss force-driven oscillation. Power has to be supplied to the system to keep the oscillation amplitude constant. This supplementary power can be derived from the middle term of (2.33) and amounts to

$$W_r = r\int_0^{t_1} v(t)^2 dt, \quad\text{and, thus}\quad P_r = r\frac{d}{dt}\int_0^{t_1} v(t)^2 dt. \quad (2.36)$$

Averaging over a full period, T, with arbitrary ϕ, finally results in

$$\overline{P} = r\frac{1}{T}\int_0^T \hat{v}^2 \cos^2(\omega t + \phi)dt = \frac{1}{2T}r\hat{v}^2 T = \frac{1}{2}r\hat{v}^2 = \frac{1}{2}\hat{F}\hat{v} = F_{\text{rms}}\,v_{\text{rms}}. \quad (2.37)$$

At the dashpot $\underline{v}$ and $\underline{F}$ are in phase, which means that the supplied power is purely resistive (active) power. This holds for the complete system when driven at its characteristic frequency. Off this frequency, additional reactive-power is needed to keep the system stationarily oscillating.

2.6 Basic Elements of Linear, Oscillating, Acoustic Systems

In addition to the mechanic elements, there is a further class of elements for oscillators that are traditionally called *acoustic elements*. Please note that the terms *mechanic* and *acoustic* are historic in this case. Since sound is mechanic, the oscillators built from both classes of elements are, to be sure, mechanic and acoustic at the same time.

The acoustic elements are formed by small cavities filled with fluid, that is, gas or liquid. In order to deal with these cavities as concentrated elements, their linear dimensions must be small compared to the wavelengths under consideration. To define the acoustic elements, we use the sound pressure, p, the sound-pressure difference, $p_\Delta = p_1 - p_2$, and the volume velocity,

$$q = \frac{dV}{dt} = A \frac{d\xi}{dt} = A\, v(t). \qquad (2.38)$$

Figure 2.7 schematically illustrates the three acoustic elements – acoustic mass, m_a, acoustic spring, n_a, and acoustic damper, r_a. Please note that here the damper and the mass are two-port elements while the spring has only one-port.

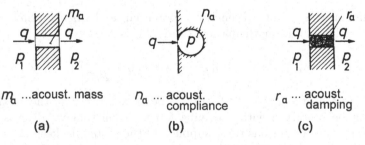

m_a ...acoust. mass n_a ... acoust. compliance r_a ... acoust. damping

(a)   (b)   (c)

Fig. 2.7. Basic elements of linear acoustic oscillators. **(a)** acoustic mass, **(b)** acoustic spring, **(c)** acoustic damper

The following equations define these elements.

• *Acoustic Mass*

$$p_\Delta(t) = m_a \frac{dq}{dt} \quad \text{or, in complex notation,} \quad \underline{p}_\Delta = j\omega\, m_a\, \underline{q} \qquad (2.39)$$

• *Acoustic Spring*

$$p(t) = \frac{1}{n_a} \int q\, dt \quad \text{or, in complex notation,} \quad \underline{p} = \frac{1}{j\omega\, n_a}\, \underline{q} \qquad (2.40)$$

• *Acoustic Damper*[5]

$$p_\Delta(t) = r_a\, q \quad \text{or, in complex notation,} \quad \underline{p}_\Delta = r_a\, \underline{q} \qquad (2.41)$$

2.7 The *Helmholtz* Resonator

The *Helmholtz* resonator is the best known example of an oscillator with an acoustic element. A *Helmholz* resonator is commonly demonstrated by blowing over the open end of a bottle to produce a musical tone. This is an auditory event with a distinct pitch that can be varied by filling the bottle with some water.

What happens when the bottle is blown on? The air in the bottle neck is a mass oscillating on the air inside the bottle, which can be considered a spring.[6]

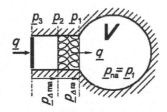

Fig. 2.8. *Helmholtz* resonator with friction

Figure 2.8 schematically illustrates the *Helmholtz* resonator with friction that causes damping. The three elements, mass, damping and spring, are connected in cascade (chain), so that the total pressure results in

$$\underline{p}_\Sigma = \underline{p}_{\Delta m_a} + \underline{p}_{\Delta r_a} + \underline{p}_{\Delta n_a}. \qquad (2.42)$$

By dividing $\underline{p}_\Sigma$ by the volume velocity, q, we arrive at the acoustic impedance, $\underline{Z}_a$, namely,

$$\underline{Z}_a = \frac{\underline{p}_\Sigma}{q} = j\omega\, m_a + r_a + \frac{1}{j\omega\, n_a}. \qquad (2.43)$$

---

[5] For the characteristic parameters of the acoustic elements, the following relations hold: $m_a = \varrho_- \, l/A$ with ϱ being density, $n_a = V/(\eta\, p_-) = V/c^2\, \varrho_-$ with $\eta = c_p/c_v$, and $r_a = \varXi\, l/A$ with $\varXi$ being flow resistivity – for details refer to Section 11.5

[6] Normally we do not experience the spring characteristics of air because the air can evacuate, but the effect in this case is similar to operating a tire pump with the opening hole pressed closed

3

Electromechanic and Electroacoustic Analogies

During the discussion of simple mechanic and acoustic oscillators in Chapter 2, readers with some electrical engineering experience may have realized that many mathematical formulae are similar to those that appear when dealing with electric oscillators. There is a general isomorphism of the equations in mechanic, acoustic and electric networks that can be exploited for describing mechanic and acoustic networks via analogous electric ones. Formulation in electrical coordinates is often to the advantage of those who are familiar with the theory of electric networks since analysis and synthesis methods from network theory can be easily and figuratively applied.

There is more than one way to portray a mechanic or acoustic network by an analogous electric one, depending on the coordinates used. To be sure, there is never a best analogy but rather one which is optimal with respect to the specific application considered. Also, please note that analogies have limits of validity. If they mimicked the problem completely, they would cease to be analogies.

For electrical engineers, dealing with mechanic and acoustic networks in terms of their electric analogies often means transforming uncommon problems into common ones, which is why they often prefer this method. Nevertheless, it is always possible to deal with the problems in their original form as well.

The following fundamental relations are to be considered when selecting coordinates for analogous representations. The two terminals of an electric circuit may serve for electric energy to be fed into the system or to be extracted from it, and in both cases the two terminals form a *port*. By restricting ourselves to monofrequent (sinusoidal) signals, it is sufficient to consider complex power instead of energy.

Electric power is the complex product of the electric voltage, $\underline{u}$, and the electric current, $\underline{i}$ – as derived in Appendix 15.2. Please note that $\underline{u}$ and $\underline{i}$ denote electric coordinates in complex notation, with peak values as magnitudes. Thus, the complex electric power is

$$\underline{P}_{el} = \frac{1}{2}\, \underline{u}\, \underline{i}^{*}, \tag{3.1}$$

with the asterisk denoting the conjugate complex form. By applying the asterisk to $\underline{i}$ and not to $\underline{u}$, we have defined inductive reactive power as positive.

The terminals of mechanic elements and the openings of acoustic elements also form ports, but in these cases, in contrast to the electrical case, one terminal or opening forms a port by itself.

The *mechanic power* is defined as the complex product of force, $\underline{F}$, and particle velocity, $\underline{v}$, as follows,

$$\underline{P}_{mech} = \frac{1}{2}\, \underline{F}\, \underline{v}^{*}. \tag{3.2}$$

Analogously, we get the *acoustic power* as the complex product of sound pressure, $\underline{p}$, and volume velocity, $\underline{q}$,

$$\underline{P}_{a} = \frac{1}{2}\, \underline{p}\, \underline{q}^{*}. \tag{3.3}$$

Since the asterisk has been applied to $\underline{v}$ and $\underline{q}$, the reactive power of mass is defined to be positive.

To arrive at isomorphisms, we use the electrical coordinates, $\underline{u}$ and $\underline{i}$, in analogy to the mechanic, $\underline{F}$ and $\underline{v}$, or the acoustic ones, $\underline{p}$ and $\underline{q}$. These analogies are restricted by the fact that the electrical coordinates are one-dimensional and can only represent one dimension of the mechanic/acoustic coordinates. For the vectors $\vec{F}$, $\vec{v}$, and $\vec{q}$, this means that only the spatial component that excites the terminal or opening in the normal direction is represented.

3.1 The Electromechanic Analogies

There are two kinds of analogies possible with mechanic networks. Analogy #1, usually called *impedance analogy*[1], is expressed as

$$\underline{F} \hat{=} \underline{u} \quad \text{and} \quad \underline{v} \hat{=} \underline{i}, \tag{3.4}$$

and analogy #2, also known as *mobile analogy* or *dynamic analogy*, is expressed as

$$\underline{F} \hat{=} \underline{i} \quad \text{and} \quad \underline{v} \hat{=} \underline{u}. \tag{3.5}$$

Both kinds of electromechanic analogies are used in praxi and shall be discussed here. Figure 3.1 provides an overview.

---

[1] The names for the analogies are traditional but may make sense in the light of the discussion in Section 3.6

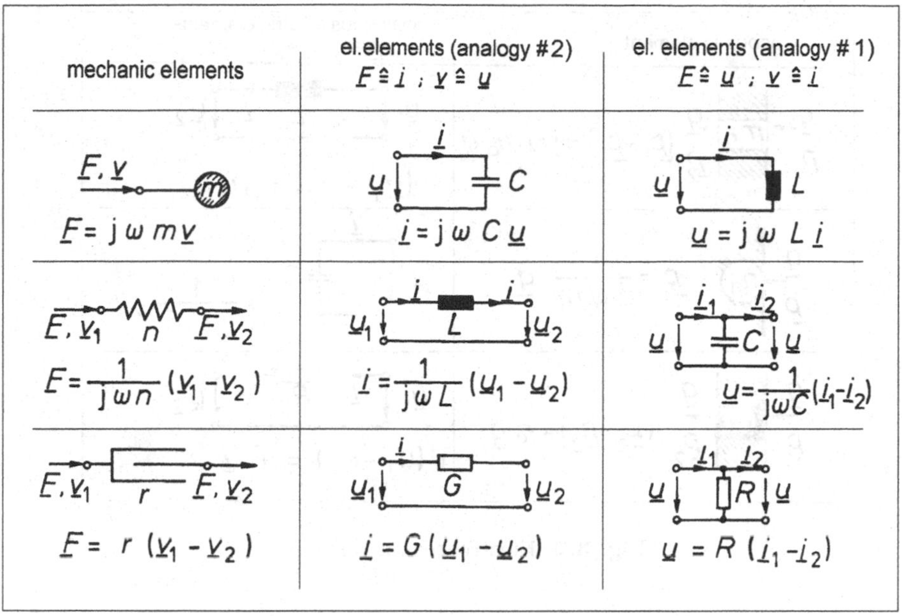

mechanic elements	el.elements (analogy #2) $F \hat{=} i \; ; \; \underline{v} \hat{=} \underline{u}$	el. elements (analogy # 1) $F \hat{=} \underline{u} \; ; \; \underline{v} \hat{=} \underline{i}$
$F = j \omega m \underline{v}$	$\underline{i} = j \omega C \underline{u}$	$\underline{u} = j \omega L \underline{i}$
$F = \dfrac{1}{j \omega n} (\underline{v}_1 - \underline{v}_2)$	$\underline{i} = \dfrac{1}{j \omega L} (\underline{u}_1 - \underline{u}_2)$	$\underline{u} = \dfrac{1}{j \omega C}(\underline{i}_1 - \underline{i}_2)$
$F = r (\underline{v}_1 - \underline{v}_2)$	$\underline{i} = G(\underline{u}_1 - \underline{u}_2)$	$\underline{u} = R (\underline{i}_1 - \underline{i}_2)$

Fig. 3.1. Electromechanic analogies

3.2 The Electroacoustic Analogy

While both variants of electromechanic analogies are used in praxi, this is not the case with the electroacoustic ones. Here only one of the two possible analogies is actually used, namely,

$$\underline{p} \hat{=} \underline{u} \quad \text{and} \quad \underline{q} \hat{=} \underline{i}. \tag{3.6}$$

Figure 3.2 presents the overview.

Please note that all analogies dealt with in Sections 3.1. and 3.2. relate to networks with lumped (concentrated) elements. This means that wave propagation is not considered. Accordingly, it is required that the acoustic elements be small compared to the wavelength of longitudinal waves across the dimensions of the elements[2]. We also assume that the individual elements are decoupled in every way except through their terminals.

3.3 Levers and Transformers

Besides m, n, r and L, C, R, respectively, there is an additional mechanic linear element that is frequently found in practical networks, namely, the me-

---

[2] An additional type of electroacoustic analogy that allows for waves will be introduced later in Section 8.5

acoustic element	analogous electric elements $\underline{p} \,\hat{=}\, \underline{u}, \; \underline{q} \,\hat{=}\, \underline{i}$
$\underset{\underline{P}_1}{\overset{\underline{q}\to}{}}\;\boxed{m_a}\;\underset{\underline{P}_2}{\overset{\to\underline{q}}{}}\;\; (\underline{p}_1 - \underline{p}_2) = j\omega m_a \underline{q}$	$\underline{u}_1 \left. \begin{array}{c} \underline{i} \;\; L \;\; \underline{i} \end{array} \right\downarrow \underline{u}_2$ $(\underline{u}_1 - \underline{u}_2) = j\omega L \underline{i}$
$\underset{\underline{p}}{\overset{\underline{q}\to}{}}\;(n_a)\;\; \underline{p} = \dfrac{1}{j\omega n_a}\,\underline{q}$	$\underline{u} \quad \overline{\overline{}}C \quad \underline{u} = \dfrac{1}{j\omega C}\,\underline{i}$
$\underset{\underline{P}_1}{\overset{\underline{q}\to}{}}\;\boxed{r_a}\;\underset{\underline{P}_2}{\overset{\underline{q}\to}{}}\;\; (\underline{p}_1 - \underline{p}_2) = r_a \underline{q}$	$\underline{u}_1 \left. \begin{array}{c} \underline{i} \;\; R \;\; \underline{i} \end{array} \right\downarrow \underline{u}_2$ $(\underline{u}_1 - \underline{u}_2) = R \underline{i}$

Fig. 3.2. Electroacoustic analogy

chanic lever. Its electric counterpart is the ideal, galvanically coupled (single-coil) transformer. Both lever and single-coil transformer are triple-port elements. Figure 3.3 illustrates the isomorphic relationships for the free-floating lever in static equilibrium for both kinds of electromechanic analogies.

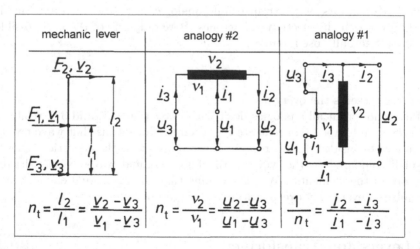

mechanic lever	analogy #2	analogy #1
$n_t = \dfrac{l_2}{l_1} = \dfrac{\underline{v}_2 - \underline{v}_3}{\underline{v}_1 - \underline{v}_3}$	$n_t = \dfrac{\underline{v}_2}{\underline{v}_1} = \dfrac{\underline{u}_2 - \underline{u}_3}{\underline{u}_1 - \underline{u}_3}$	$\dfrac{1}{n_t} = \dfrac{\underline{i}_2 - \underline{i}_3}{\underline{i}_1 - \underline{i}_3}$

Fig. 3.3. Ideal one-coil transformers as electric analogies for the free-floating mechanic lever. l ...lever length, ν ...number of turns, n_t ...transformation ratio

In the domain of electro-acoustic analogies, a lever does not exist. The so-called *velocity transformer* – sketched in Fig. 3.4 – is frequently mistaken for an acoustic lever, but it actually acts as a mechanic lever with one terminal fixed to ground.

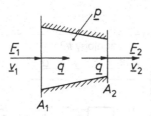

Fig. 3.4. Velocity transformer

Please note that mass and compliance in the cavity are neglected and that the linear dimensions of the cone are small compared to the wavelength. With $\underline{q}$ and $\underline{p}$ being continuous, one gets

$$\underline{q} = \underline{v}_1 A_1 = \underline{v}_2 A_2 \quad \text{and} \quad \underline{p} = \frac{F_1}{A_1} = \frac{F_2}{A_2} . \tag{3.7}$$

Introducing $n_t = A_1/A_2$ for the area ratio leads to the following equations,

$$\underline{v}_2 = n_t \underline{v}_1 \quad \text{and} \quad \underline{F}_2 = \frac{1}{n_t} \underline{F}_1 , \tag{3.8}$$

and, consequently,

$$\underline{Z}_{2,\,\text{mech}} = \frac{1}{n_t^2} \underline{Z}_{1,\,\text{mech}} . \tag{3.9}$$

Velocity transformers are applied as *impedance transformers* as exemplified by the compression chamber at the mouth of a horn loudspeaker – see Section 5.2 for details.

3.4 Rules for Deriving Analogous Electric Circuits

When deriving the analogous electric circuit of a mechanic or acoustic circuit, the mechanic or acoustic one-, two- or triple-port elements must be replaced by analogous electric elements. When connecting those elements, the following rules apply.

For electromechanic analogies – refer to Fig. 3.5,

- Chains (cascades) of mechanic elements result in chains of electric elements. The masses or their analogous single-port electric elements always form the end of a chain or of a branch

- Branching of mechanic two-port elements leads to parallel branching in analogy # 2 and to serial branching in analogy # 1, in each case by means of rigid, massless rods. Again, the single-port elements form the end elements

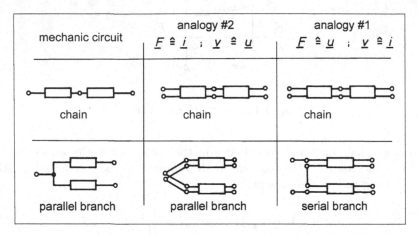

Fig. 3.5. Electromechanic analogies for mono-, dual- and triple-port elements

For electroacoustic analogies,

- Chains of acoustic elements result in chains of electric elements. The single-port spring and its analogous electric element form end elements

- Parallel branching of acoustic elements leads to parallel branching of electric elements

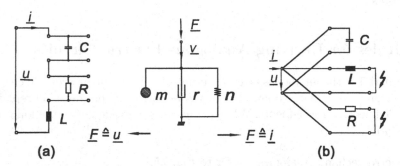

Fig. 3.6. Electric analogies of a simple mechanic parallel-branch oscillator, (a) analogy # 1, (b) analogy # 2

3.5 Synopsis of Electric Analogies of Simple Oscillators

The schematic in Fig. 3.6 shows how the electric analogies are derived for mechanic *parallel-branch oscillators*, often simply called parallel oscillators, and further, how the electric analogies are derived for mechanic *serial-branch oscillators*, often simply called serial oscillators. Figure 3.7 provides a synopsis of the different possible analogous relationships.

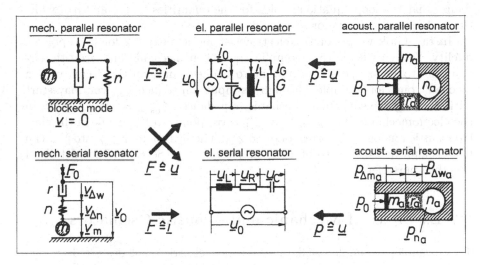

Fig. 3.7. Synopsis of the electric analogies of simple mechanic and acoustic oscillators

3.6 Circuit Fidelity, Impedance Fidelity and Duality

By looking at the electromechanic analogies given in Fig. 3.7, it becomes apparent that the circuits derived by analogy #2, namely, with $\underline{F} \mathrel{\hat{=}} \underline{i}$ and $\underline{v} \mathrel{\hat{=}} \underline{u}$, show the same topology as their mechanic counterpart. This behavior is called *circuit fidelity* or topological fidelity. Please note that in this case impedances transform into admittances and vice versa.

However, those circuits derived with analogy #1, that is, with $\underline{F} \mathrel{\hat{=}} \underline{u}$ and $\underline{v} \mathrel{\hat{=}} \underline{i}$, result in a topology that is *dual* with respect to the mechanic original. In this case the impedances lead to isomorphic expressions, what is called *impedance fidelity*. The circuit topologies transform into the dual ones.

In electrical networking terminology, the term *dual* refers to two circuits where one behaves in terms of voltages just as the other one behaves in terms of currents. We find that $\underline{Y} = \text{const}^2 \, \underline{Z}$ for the elements of dual circuits. This means that the impedance of the one circuit is proportional via a real constant to the admittance of its dual pair. Important dual pairs include the elements capacitance, C, v.s. inductance, L, and resistance, R, v.s. conductance, G. Further, in dual networks, closed loops of one (meshes) correspond to nodes of the other, and vice versa. Consequently T-circuits correspond to π-circuits, and serial-branching circuits correspond to parallel-branching ones.

The electroacoustic analogy that we use possesses both circuit fidelity and impedance fidelity, which is why the other possible analogy is never applied.

To understand the characteristic features discussed above, it is helpful to realize that the loop equation[3] holds for the quantities $\underline{u}$, $\underline{v}_\Delta$ and $\underline{p}_\Delta$, while the node equation holds for $\underline{i}$, $\underline{F}$, and $\underline{q}$.

In this book we prefer the electromechanic analogy $\#\,2$ for its topologic fidelity. Yet, this leads to a complication when mechanic and acoustic circuits are to be merged. If you want to connect an acoustic circuit with a mechanic one, for example, through its input impedance $\underline{Z}_a$, you may start with deriving the equivalent mechanic impedance, $\underline{Z}_{\mathrm{mech}} = A^2\,\underline{Z}_a$. Now, in the electromechanic analogy $\#\,2$, $\underline{Z}_{\mathrm{mech}}$ corresponds to $\underline{Y}_{\mathrm{el}}$, while in the electroacoustic analogy $\underline{Z}_a$ corresponds to $\underline{Z}_{\mathrm{el}}$. The inversion of $\underline{Z}_{\mathrm{el}}$ into $\underline{Y}_{\mathrm{el}}$ can be accomplished by means of an ideal *gyrator* – see Section 4.3 for details of this dual-port element.

3.7 Examples of Mechanic and Acoustic Oscillators

Two examples of simple oscillators and their electric analogies are described below. The first is a *mechanic oscillator* with two finite masses. This kind of oscillator can be found in many practical applications, including engines dynamically based on concrete plates, ultrasound-source transducers, and vibrating engine parts. The circuit diagrams are given in Fig. 3.8.

The characteristic frequency for the mechanic oscillator is

$$\omega_0 = \frac{1}{\sqrt{n\,m_\Sigma}}. \tag{3.10}$$

This relationship becomes evident by looking at the analogue electric circuit and noting that the two capacitances are serially linked. Consequently, the effective mass is

$$m_\Sigma = \frac{m_1\,m_2}{m_1 + m_2}. \tag{3.11}$$

Figure 3.9 shows a simple *cavity resonator* with two finite compliances, otherwise known as an *acoustic oscillator*. Such closed-cavity resonators are, for example, applied for calibration of microphones because they are well insulated

---

[3] Recall that in electrical terms the loop equation is $\sum \underline{u}_\mathrm{n} = 0$, with $n = 1, 2, 3, \cdots$, meaning that by completely circling a mesh we end at the same electric potential. The node equation is $\sum \underline{i}_\mathrm{n} = 0$, with $n = 1, 2, 3, \cdots$, meaning that all electric charge that flows into a node must leave it at the same time

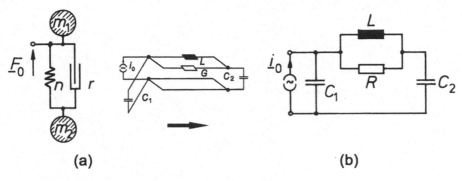

(a)   **(b)**

Fig. 3.8. Analogous circuit diagram (analogy # 2) of a two-mass mechanic oscillator, (a) mechanic arrangement, (b) analogous electric circuit. The plot in the middle illustrates how the electric circuit is derived

from external noise and have low power losses as may be caused by radiation. A vibrating piston excites the system and delivers an exactly known volume velocity of q_0.

The characteristic frequency of the acoustic resonator is

$$\omega_0 = \frac{1}{\sqrt{m_a n_{a\Sigma}}}, \tag{3.12}$$

and its effective compliance is

$$n_{a\Sigma} = \frac{n_{a_1} n_{a_2}}{n_{a_1} + n_{a_2}}. \tag{3.13}$$

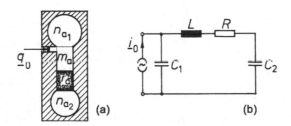

(a)   **(b)**

Fig. 3.9. Circuit diagram for a two-cavity acoustic oscillator, (a) acoustic arrangement, (b) analogous electric circuit

4

Electromechanic
and Electroacoustic Transduction

In the preceding chapter, we dealt with simple linear, time-invariant mechanic and acoustic networks and their electric analogies. In general, these networks can be quite complicated and may assume any number of degrees of freedom. Yet, regardless of how sophisticated the networks are, the energy and power transported in these networks is either mechanic, acoustic, or electromagnetic. Acoustic power and energy are of mechanic nature. Thus the terminological distinction between mechanical, F, v, and acoustical coordinates, p, q, is purely operational. In this chapter, we shall present the possibility of coupling electrical and mechanical domains, which results in a coupling of electric and mechanic energy and power. This topic is extremely important for modern acoustics.

This coupling can be manifold. The coupling element, the *electromechanic coupler*, can contain its own power sources and may be either active or passive. The relationship between mechanic/acoustic and electric coordinates can be linear or nonlinear. Coupling may be bi-directional, that is, exist for both directions, electric to mechanic and vice versa, or only mono-directional. It may be retroactive or not.

In this chapter we restrict ourselves to examples of practical importance. On the mechanic/acoustic side we use the coordinates F and v, which can be transformed into p and q, given that the effective area, $A_\perp$, is known. In most practical cases, a linear and time-invariant physical relationship between F, v and u, i is presumed. If this is not the case, approximate linearity may be assumed for small alternating quantities superimposed on large steady offsets.

4.1 Electromechanic Couplers
as Two- or Three-Port Elements

A coupling element between mechanic and electric domains is generally represented as a three-port representation – shown in Fig. 4.1 (a). In a housing of mass, m, rigidly connected to the terminal 2, there is a *movable component* which can be operated by means of a rigid, massless rod. This rod penetrates the housing from the left, denoted terminal 1 in the figure.

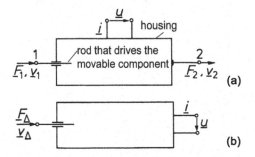

Fig. 4.1. Black-box representations of a coupling element, **(a)** three-port representation, **(b)** two-port representation

The power that is transported into the movable component is

$$\underline{P}_\Delta = \frac{1}{2}\,\underline{F}_\Delta\,\underline{v}_\Delta^*\,. \tag{4.1}$$

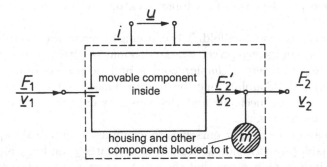

Fig. 4.2. Schematic representation of a coupling element, distinguishing mobile components and components blocked to the housing

The schema shown in Fig. 4.2 is the result of counting all masses rigidly blocked to the housing as part of the housing mass, m. Assuming lossless coupling at this point we have the following balance of power,

$$\underline{F}_\Delta \, \underline{v}_\Delta^* = \underline{F}_1 \, \underline{v}_1^* - \underline{F}_2' \, \underline{v}_2^* \,, \tag{4.2}$$

which consequently leads to

$$\underline{F}_\Delta = \frac{\underline{F}_1 \, \underline{v}_1^* - \underline{F}_2' \, \underline{v}_2^*}{\underline{v}_1^* - \underline{v}_2^*} \quad \text{with} \quad \underline{v}_\Delta = \underline{v}_1 - \underline{v}_2 \,. \tag{4.3}$$

When the housing is fixed to the ground, as is frequently the case, $\underline{v}_2$ becomes zero and yields $\underline{v}_\Delta = \underline{v}_1$ and $\underline{F}_\Delta = \underline{F}_1$. In this case, terminal 2 may be disregarded because no power passes through it.

Regardless of the specific case, the essential role of the electromechanic coupler is to couple the introduced mechanic power, $(1/2)\,\underline{F}_\Delta \underline{v}_\Delta^*$, at one port and the provided electric power, $(1/2)\,\underline{u}\,\underline{i}^*$, at the other – or vice versa. Figure 4.1 (b) illustrates the situation in the form of a two-port element. This figure will be the basis for the rest of this chapter.

4.2 The Carbon Microphone – A Controlled Coupler

An important class of electromechanic couplers consists of couplers where the signal-representing quantities in one network, mechanic or electric, accomplish the coupling by controlling elements of the other network.

An illustrative historic example is the carbon microphone, which was an important part of telephone technology for about a century and was, during that period, the most frequently-used microphone type worldwide.

The carbon microphone is a unidirectional coupler working in the mechanic to electric direction. It requires a DC power supply and behaves as an active element at its output port. It can actually perform a power amplification on the order of 30, which is about 15 dB. This property was the main reason for its widespread use at a time when telephone terminals had no other built-in amplifiers.

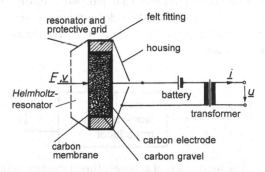

Fig. 4.3. Section view of a carbon microphone

Figure 4.3 illustrates a section through a carbon microphone. The electrically conducting (carbone) membrane is excited by the sound-pressure impinging on it.

A carbon electrode is positioned behind the membrane, and the gap between this back electrode and the membrane is filled with fine-grain carbon gravel. Further, a *Helmholtz* resonator may be arranged in front of the membrane to boost the sensitivity in the main spectral range of speech signals.

The electric resistance of the gravel, about $100\,\Omega$, varies according to the alternating pressure on it. If a DC current is applied to this arrangement, the current will be superimposed on an alternating current as the resistor varies. The AC component of the current can be filtered out using a transformer – shown in Fig. 4.3. Typical carbon microphones have a sensitivity of $T_{\mathrm{up}} \approx 500\,\mathrm{mV/Pa}$, which corresponds to 500 mV for a 94-dB sound.

Several disadvantages should be mentioned though. First, carbone microphones produce many upper harmonics with a power of up to 25% of the fundamental harmonic. Second, they generate quite a bit of internal noise, and, finally, they are power consuming. For these reasons, these microphones have been replaced in modern technology with electret or chip microphones – see Section 6.6.

Other examples of controlled couplers are the foil-strain gauge (a controlled resistor), the piezotransistor (a pressure-sensitive transistor), and the compressed-air loudspeaker in which an electromagnetic valve controls a stream of compressed air to generate sounds of up to 160 dB.

4.3 Fundamental Equations of Electroacoustic Transducers

From an application standpoint, the most important electromechanic couplers are those where electric power is directly transformed into mechanic power, or vice versa. This class of coupler is called *transducers*. The term transducer is usually reserved for those couplers that can work bi-directionally and are intrinsically passive, meaning that they do not have power sources of their own. Coupling by means of transducers is retroactive because there is power flowing across the domains.

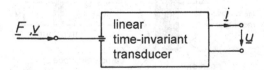

Fig. 4.4. Schematic plot of a linear, time-invariant transducer

The following system of linear equations applies when the transducers are linear and time-invariant. This most important case is schematically illustrated in Fig. 4.4.

$$\begin{pmatrix} \underline{v} \\ \underline{F} \end{pmatrix} = \begin{pmatrix} \underline{A}_{11} & \underline{A}_{12} \\ \underline{A}_{21} & \underline{A}_{22} \end{pmatrix} \begin{pmatrix} \underline{u} \\ \underline{i} \end{pmatrix}, \tag{4.4}$$

where we omit the subscript Δ from now on for simplicity. This form of the fundamental transducer equations is called the primary form. The coefficients of the transfer matrix, A_{ik}, are called chain parameters.

We will now concentrate on transducers that transform power using force effects in electromagnetic fields. In such transducers the *Lorentz* force is effective. It can be generally expressed as

$$\vec{F} = Q_{el}\,\vec{E} + Q_{el}\,(\vec{v} \times \vec{B}), \tag{4.5}$$

or, in expanded form, with l being the path of i in the B-field,

$$\vec{F} = (C\,u)\vec{E} + i\,(\vec{l} \times \vec{B}). \tag{4.6}$$

If we assume the simple case in which the electric field strength, $\vec{E}$, and the magnetic-flux density, $\vec{B}$, are constant, we can derive the following principles from these equations for the forces in transducers[1].

- For purely electric fields the force, $\underline{F}$, is proportional to the voltage, $\underline{u}$
- For purely magnetic fields the force, $\underline{F}$, is proportional to the current, $\underline{i}$

Real transducers have additional components beyond the actual energy-transducing elements. Transducers working with magnetic fields always contain an inductance, and those working with electric fields always have a capacitance. We also have to expect a resistance, representing electric power losses. On the mechanical side, mass is unavoidable. A spring is required to provide a restoring force on the oscillating mass and to compensate for gravitation, and there is usually some mechanic damping as well.

It is possible to mathematically separate the real transducer into a chain of three two-port elements – depicted in Fig. 4.5. The *inner transducer* can be configured as ideal, that is, without any resistances, dampers, or reactances. The inner transducer cannot store energy in any way because it is lossless, and it is not directly accessible from the outside. For the complex power at its ports we can write

$$\underline{F}_i\,\underline{v}_i^* = \underline{u}_i\,\underline{i}_i^*. \tag{4.7}$$

We shall now derive the fundamental equations for inner transducers based on either magnetic or electric fields.

---

[1] In general, higher forces can be achieved with magnetic fields because the electric field-strength is limited by the danger of disruptive discharge. This is the reason that electric motors and generators usually use magnetic fields

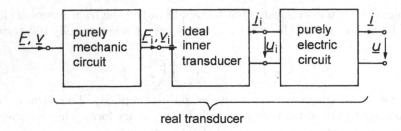

Fig. 4.5. A real transducer as a chain of three two-ports

As noted above, $\underline{F}$ and $\underline{v}$ are proportional in magnetic-field transducers. The proportionality coefficient, M, is a real constant and specific to a particular transducer. With the movable component blocked, meaning $\underline{v}_i = 0$, and the electric port cut short, meaning $\underline{u}_i = 0$, we measure a force, $\underline{F}_i$, when applying a current, $\underline{i}_i$, as follows,

$$\underline{F}_i = M \, \underline{i}_i. \tag{4.8}$$

Since energy is neither lost nor stored in the inner transducer, the power at the two ports is identical by definition. Consequently, we also know the relationship between $\underline{v}_i$ and $\underline{u}_i$ when $\underline{F}_i$ and $\underline{i}_i$ are zero, namely,

$$\underline{v}_i = \frac{1}{M} \, \underline{u}_i. \tag{4.9}$$

A combination of the two yields

$$\begin{pmatrix} \underline{v}_i \\ \underline{F}_i \end{pmatrix} = \begin{pmatrix} 1/M & 0 \\ 0 & M \end{pmatrix} \begin{pmatrix} \underline{u}_i \\ \underline{i}_i \end{pmatrix}. \tag{4.10}$$

By applying an electromechanic analogy #2, we can identify the ideal transformer as analogy of the magnetic transducer shown in Fig. 4.6.

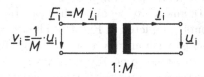

Fig. 4.6. The ideal transformer as analogy for the ideal magnetic-field transducer

Electric-field transducers show proportionality of $\underline{F}$ and $\underline{u}$. The proportionality coefficient, N, is again a real constant, and specific to a particular transducer. With the movable component fixed so that $v_i = 0$, the equation is

$$\underline{F}_i = N \, \underline{u}_i, \tag{4.11}$$

and, due to the identity of the power at the two ports, it follows that

$$\underline{v}_i = \frac{1}{N}\, \underline{i}_i,$$ (4.12)

which results in the subsequent matrix equations,

$$\begin{pmatrix} \underline{v}_i \\ \underline{F}_i \end{pmatrix} = \begin{pmatrix} 0 & 1/N \\ N & 0 \end{pmatrix} \begin{pmatrix} \underline{u}_i \\ \underline{i}_i \end{pmatrix}.$$ (4.13)

Electromechanic analogy #2 renders the ideal gyrator as analogy for the electric-field transducer – see Fig. 4.7. The ideal gyrator is a two-port element that is dual to the ideal transformer[2].

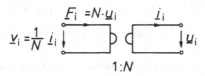

Fig. 4.7. The ideal gyrator as analogy of the ideal electric-field transducer

4.4 Reversibility

In mechanics as well as in electric networks, the principle of reciprocity may apply. In mechanics, for example, we can observe it in experiments like the one shown in Fig. 4.8.

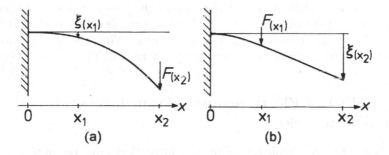

Fig. 4.8. Mechanic reciprocity experiment with a bending beam, **(a)** Force applied at position x_2, deflection at position x_1, **(b)** vice versa

The experiment demonstrates that when we apply a force to a bending beam at position 2 and observe a deflection of the beam in position 1, the ratio of

---

[2] When using the analogy #1, the transformer would represent the electric-field transducer and the gyrator the magnetic-field tranducer

force and deflection is the same as would be observed if the force were applied at position 1 and the deflection were observed at position 2, assuming that all other forces are zero. This situation is illustrated by the following equation.

$$\left.\frac{F(x_2)}{\xi(x_1)}\right|_{F(x_1)=0} = \left.\frac{F(x_1)}{\xi(x_2)}\right|_{F(x_2)=0}. \tag{4.14}$$

The same principle holds in electric networks for the ratio of voltages at one port and currents at another, assuming that these networks are only constructed of inductances, capacitances, resistances, and ideal transformers. Mathematically, the formulation of the principle of reciprocity is as follows,

$$\det|\underline{A}| = \underline{A}_{11}\,\underline{A}_{22} - \underline{A}_{12}\,\underline{A}_{21} = +1\,, \tag{4.15}$$

with chain parameters and by using the chain reference system for currents and voltages – see Fig. 4.4.

In order to cover electric-field transducers, gyrators must be included as additional elements, which requires us to modify equation (4.14). This requirement becomes clear when we use analogy #2 to check the reciprocity for magnetic-field and electric-field transducers. For magnetic-field transducers we find $\frac{1}{M}\,M - 0 = +1$, but for electric-field transducers, $0 - \frac{1}{N}\,N = -1$. Compare Figs. 4.6 and 4.7 for more clarification. The gyrator obviously introduces a 180° phase shift, but this phase shift is irrelevant in most cases. Please note that it can be introduced just by the choice of analogy!

Consequently, it is sufficient to require only that $\det|\mathcal{A}| \overset{!}{=} 1$, which leads to the following, more general equation,

$$\left.\left|\frac{u}{F}\right|\right|_{i=0} = \left.\left|\frac{v}{i}\right|\right|_{F=0}. \tag{4.16}$$

This equation essentially says that the ratio of the power at port 1 and port 2 is independent of the direction of transduction, provided that both ports are terminated with a real, purely resistive impedance. When the above equation holds, we call a transducer *power symmetric* or *reversible*.

4.5 Coupling of Electroacoustic Transducers to the Sound Field

In order for electromechanic transducers to work, they must be coupled to the sound field in such a way that they can either act as *receivers*[3] by withdrawing power from the field or as *emitters*[4] by delivering power to it. Fig. 4.9 illustrates these roles. In this way, we have moved from electromechanic to electroacoustic transducers.

---

[3] Note that in hearing-aid technology the sound emitter is called receiver because it receives electrical signals

[4] Emitters are sometimes also called *transmitters*, to denote that an electrical signals is transmitted to the sound field

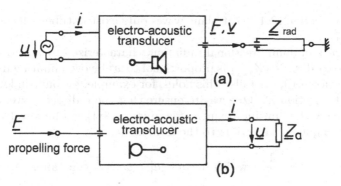

Fig. 4.9. Coupling electromechanic transducers to the sound field renders electro-acoustic transducers

Let us consider the sound emitters first. The following thought experiment may help us to better understand sound-field coupling at the emitter's end – see Fig. 4.10. A sound emitter may be operated in a vacuum. In this case, coupling does not take place because there is no sound field. The mechanic output is idle, meaning that $F = 0$. An electric input impedance of $\underline{Z}_{\rm el}|_{F=0} = \underline{u}/\underline{i}$ can be measured at the input of the transducer in this condition.

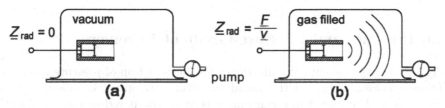

Fig. 4.10. Thought experiment to illustrate the radiation impedance, **(a)** evacuated volume, no sound radiation, **(b)** gas-filled volume, sound is radiated

Now let air flow into the volume. A different input impedance is measured afterwards, which is explained by the fact that the emitter is no longer idle at the output port. This port is now terminated by a finite impedance called the *radiation impedance*, $\underline{Z}_{\rm rad} = \underline{F}/\underline{v}$. The radiation impedance is a mechanic impedance. The emitter now delivers power to the sound field that can be expressed as

$$\overline{P} = \frac{1}{2}\,\mathrm{Re}\,\{\underline{Z}_{\rm rad}\}\,|\underline{v}^2| = \frac{1}{2}\,r_{\rm rad}\,|\underline{v}|^2 = \frac{1}{2}\,\mathrm{Re}\,\{\underline{F}\,\underline{v}^*\}. \qquad (4.17)$$

The radiation impedance depends on the type of sound field. In the case of plane waves, it is a real quantity. We than have $\underline{Z}_{\rm rad} = r_{\rm rad}$, where $r_{\rm rad}$ is called *radiation resistance*. In the free field, plane waves are hard to realize, but they may be approximated in the beam of a highly directional sound source – see Section 10.4 Exact plane waves only exist in tubes – as dealt with in Section 7.5.

In the case of omnidirectional radiation, characterized by spherical sound fields of zero order, $\mathrm{Re}\,\{\underline{Z}_{\mathrm{rad}}\}$ is proportional to ω^2 below a limiting frequency, ω' – see Section 9.1 for details. This holds, for example, for built-in loudspeakers – refer to Section 5.2. One way to ensure that the radiated power does not decrease below the limiting frequency is to increase the volume velocity with decreasing frequency according to the formula

$$|v|^2 \sim \frac{1}{\omega^2}\,,\ \text{which means}\ |a| = \omega\,|v| \stackrel{!}{=} \text{constant}\,. \tag{4.18}$$

For sound receivers, the coupling to the sound field depends on directional characteristics as well. The relevant question in this case is which sound-field quantity drives the movable component of the transducer because this determines the receiver principle. The most important cases are the following.

- The driving force is proportional to the sound pressure, that is $F \sim p$
- The driving force is proportional to the sound-pressure gradient, namely, $F \sim \partial p/\partial r$

We speak of *pressure receivers* in the former and *pressure-gradient receivers* in the latter case.

4.6 Pressure and Pressure-Gradient Receivers

Figure 4.11 (a) schematically illustrates the construction of pressure receivers. There is a closed volume with a membrane of effective area, A, covering part of the surface. The complete arrangement is small compared to the wavelength of the sound, λ.

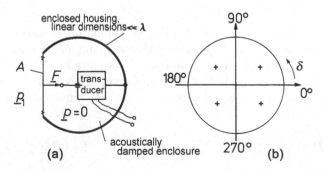

Fig. 4.11. Pressure receiver, (a) construction, (b) directional characteristic

The driving force in this case turns out to be $\underline{F} = A\underline{p}_1$. This relation does not depend on frequency, that is, $\underline{F} \neq f(\omega)$. The sensitivity of the device has a spherical directional characteristic[5], $\varGamma = 1$ – see in Fig. 4.10 (b). Here we take the microphone axis as the reference direction. In other words, $\varGamma = \underline{F}(\delta)/\underline{F}_{\max}$, with δ being the angle between microphone axis and the sound-incidence direction. Please note for all plots of directional characteristics that they have to be considered as three-dimensional, although only the vertical projection is shown here.

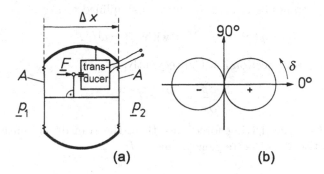

(a)   **(b)**

Fig. 4.12. Pressure-gradient receiver, **(a)** construction, **(b)** directional characteristic

Figure 4.12 (a) illustrates the construction of pressure-gradient receivers. As with the pressure receivers, the linear dimensions are small compared to the wavelength. There are two membranes, one on each side of the otherwise closed volume, and each having effective area A. The movable component is coupled to the membrane in such a way that its driving force is equal to the difference of the forces affecting each membrane. The same can, by the way, be achieved with only one membrane that is accessible for sound from both sides. The driving force, consequently, becomes

$$\underline{F} = A\left(\underline{p}_1 - \underline{p}_2\right) = A\,\underline{p}_\Delta .$$ (4.19)

For the pressure difference between two points in a sound field in the direction of wave propagation, we have

$$\underline{p}_\Delta = \frac{\partial p}{\partial x}\,\Delta x ,$$ (4.20)

with $\partial p/\partial x$ being a vector, called sound-pressure gradient, $\overrightarrow{\mathrm{grad}\,p}$. To create a pressure difference between the two membranes, only that portion of $\overrightarrow{\mathrm{grad}\,p}$ that coincides with the microphone axis becomes effective. This portion is

---

[5] $\varGamma$, the directional characteristic, is the ratio of the magnitude of the driving force taken for a sound incidence from a certain direction compared to the magnitude of the driving force in the direction of maximum sensitivity

$$\underline{p}_\triangle = \overrightarrow{\mathrm{grad}\, \underline{p}}\, \Delta x\, \cos\delta\,. \tag{4.21}$$

The directional characteristic – depicted in Fig. 4.12 (b) – turns out to be

$$\Gamma = \cos\delta\,. \tag{4.22}$$

This is called the *figure-of-eight* characteristic. The plus signs and minus signs in the plot denote a 0°- or 180°-phase difference, respectively, between the pressure-gradient signal and the electric output signal.

Let us now consider two special cases, a plane sound field and a spherical one. For a diverging plane wave, we have in simplified form

$$\underline{p}(x) \sim \mathrm{e}^{-\mathrm{j}\beta x}, \quad \text{with} \quad \beta = \omega/c\,, \tag{4.23}$$

as will be derived in Section 7.3. Consequently, we get

$$\frac{\partial \underline{p}}{\partial x} \sim \frac{\omega}{c}\,\mathrm{e}^{-\mathrm{j}\beta x}\,. \tag{4.24}$$

This means that the driving force and, thus, the transducer sensitivity becomes proportional to the frequency, that is $\underline{F} \sim \omega$.

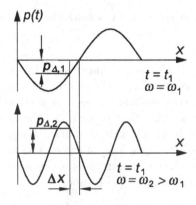

Fig. 4.13. Two sound-pressure waves of different frequencies

Figure 4.13 illustrates this fact. It shows two sound-pressure waves of different frequencies, $\omega_2 > \omega_1$, frozen at an instant $t = t_1$. Clearly, the pressure difference, $\underline{p}_\triangle$, between two points Δx apart is higher for the higher frequency. Note what would happen if the microphones were not small compared to the wavelength, λ. For a finite Δx, the pressure difference, $\underline{p}_\triangle$, would vanish for $\Delta x = n\,\lambda/2$. For real microphones this happens above about 4–10 kHz.

For a diverging spherical wave of zero order, which we shall introduce in more detail in Section 9.1, a simplified equation for the sound pressure is

$$\underline{p}(r) \sim \frac{1}{r}\,\mathrm{e}^{-\mathrm{j}\beta r}\,. \tag{4.25}$$

The factor $1/r$ is due to lossless spherical expansion, where it is assumed that the same active power passes through all spherical areas (shells) around the sound source. The sound-pressure gradient, then, results as

$$\frac{\partial p}{\partial r} \sim \left(\frac{1}{r} + \frac{j\omega}{c}\right) \frac{1}{r} e^{-j\beta r} \quad \text{with} \quad \beta = \omega/c. \tag{4.26}$$

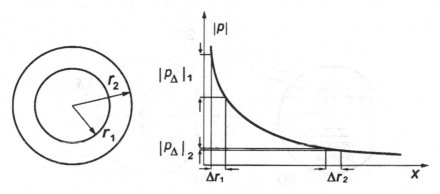

Fig. 4.14. Illustrating the near-field component of the sound-pressure gradient

When approaching the sound source, the frequency-independent sum term, $1/r$, increases relative to the frequency dependent one, $j\omega/c$. This leads to a relative gain of the low frequencies compared to the higher ones.

The first term in the sum is sometimes called the *near-field gradient*, and the second one the *far-field gradient*. While the far-field gradient originates from phase differences – explained in Fig. 4.13 – the near-field gradient stems from the decrease of amplitude with distance, an effect that is independent of frequency. Figure 4.14 further illustrates this latter effect.

Pressure-gradient receivers are usually much less sensitive than pressure receivers since $\partial p/\partial r \ll p(r)$. The relative increase of the low frequencies in the near field is exploited to construct microphones for acoustically adverse conditions like very noisy or reverberant situations. These microphones are less sensitive for distant sources than they are, for instance, for a speaker's voice when held close to the mouth. We can find them used by bus-drivers or announcers at fairs for example.

4.7 Further Directional Characteristics

When linearly superimposing a pressure and pressure-gradient receiver, one arrives at a directional characteristic called *cardioid*. Figure 4.15 (a) shows such a characteristic. The mathematical expression for it is

$$\Gamma = \frac{1}{2} \left(1 + \cos \delta \right), \tag{4.27}$$

where the reference direction for normalization is again the receiver axis. There are two ways of realizing such a receiver. The first one is to arrange both receivers in practically the same position (coincidence microphones) and add their output signals in phase and with the same amplitude at frontal incidence. The second one – illustrated in Fig. 4.15 (b) – uses an acoustic delay line to guide sound from the rear side of the receiver to the back of the membrane.

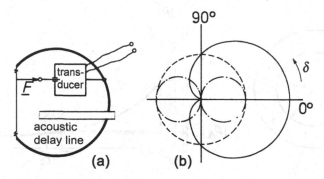

Fig. 4.15. Cardiod microphone, construction and directional characteristics

Receivers that select higher-order pressure gradients from the sound field than gradient receivers with figure-of-eight characteristics achieve even sharper spatial selectivity. The higher orders are determined according to $\partial^n p / \partial r^n$ and lead to directional characteristics as depicted in Fig. 4.16. These receivers are, however, even less sensitive and very frequency dependent, according to $\underline{F} \sim \omega^n$. The analytical expression for their directional characteristics is

$$\Gamma = \cos^n \delta. \tag{4.28}$$

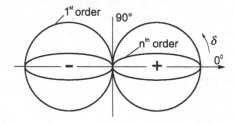

Fig. 4.16. Higher-order figure-of-eight directional characteristics

The receivers we have dealt with so far have all been small compared to the wavelengths considered, so that no interference takes place. We shall now discuss an example of a receiver that is clearly larger than typical wavelengths.

This so-called *line microphone* – schematically depicted in Fig. 4.17 – deliberately exploits interference of incoming waves.

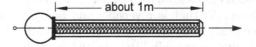

Fig. 4.17. Line microphone

The line microphone is a receiver with an extremely narrow directionality that can be realized for a fairly broad frequency band. It consists of a tube that is typically open at one end and fitted with a microphone at the other end. Along the tube there is a slit through which sound can enter.

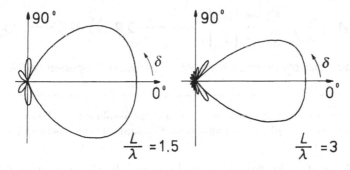

Fig. 4.18. Directional characteristics of line microphones of different length

Now consider the following two extreme cases.

- Sound impinges laterally, meaning that all points on the slit are excited in phase, prompting waves that all propagate to the microphone. They cancel each other out upon arrival because of their different travel distances, making the receiver insensitive to this direction

- Sound impinges frontally. Now all waves propagate in phase to the microphone and add up there upon arrival. The receiver has its maximum sensitivity in this direction

Figure 4.18 illustrates the complete directional characteristics for two receivers of different length. The computation of such diagrams will be discussed in Section 9.4. It should be noted that the slit is often covered with fabric with a flow resistance that varies along the tube. This compensates for losses during wave propagation in the tube.

4.8 Absolute Calibration of Transducers

This section explains how the principle of reversibility can be used for absolute calibration of an electroacoustic coupler, M. A necessary tool is a small supporting transducer, X, that is reversible and has a spherical directional characteristic. A constant sound source is also required. At the supporting transducer we have, due to reversibility as stated in (4.16),

$$\left|\frac{u}{F}\right|_{i=0} = \left|\frac{v}{i}\right|_{F=0} \quad \text{and, thus,} \quad |\underline{T}_{\mathrm{up}}|_X = \left|\frac{u}{p}\right|_{i=0} = \left|\frac{q}{i}\right|_{F=0}. \tag{4.29}$$

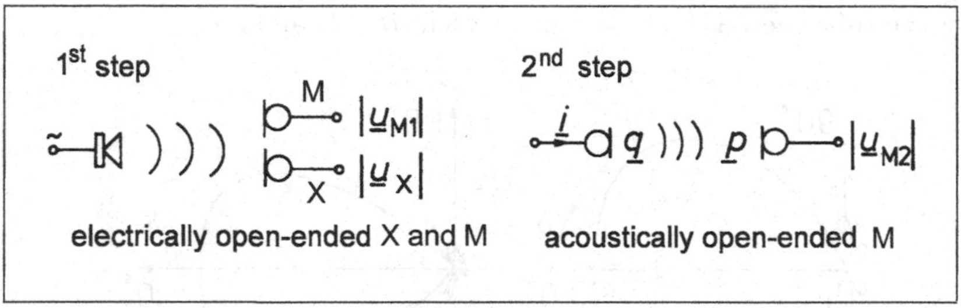

Fig. 4.19. Overview of the reversibility method for absolute transducer calibration. M ... electroacoustic coupler to be calibrated, X ... supporting transducer (reversible)

Figure 4.19 denotes the two steps to be taken. In the first step, both the supporting transducer and the microphone to be measured are positioned closely together (spatially coincident) in a free sound field. The output voltages of both components are measured with their electric output ports at idle. The ratio of the two voltages measured corresponds to the ratio of the sensitivities of the two transducers and is expressed as

$$\left|\frac{u_X}{u_{M_1}}\right| = \frac{|\underline{T}_{\mathrm{up}}|_X}{|\underline{T}_{\mathrm{up}}|_M}. \tag{4.30}$$

In a second step, we now use the supporting transducer as a spherical sound source. This source is excited with a current, i, and generates a volume velocity of q. Due to the negligible power efficiency of such a source, we can assume that its acoustic port is mechanically idling, that is $F = 0$. This means with (4.29) that we have

$$|q| = |\underline{T}_{\mathrm{up}}|_X\, |i|. \tag{4.31}$$

With the acoustic impedance, $|p/q| = |\underline{Z}_a|$, which can be computed for the spherical sound field with the distance known, we arrive at

$$|u|_{M_2} = |p|\, |\underline{T}_{\mathrm{up}}|_M = |q|\, |\underline{Z}_a|\, |\underline{T}_{\mathrm{up}}|_M. \tag{4.32}$$

Inserting (4.30) into (4.31) produces

$$|\underline{q}| = |\underline{T}_{up}|x\,|i| = \left|\frac{u_X}{u_{M_1}}\right|\,|\underline{T}_{up}|_M\,|\underline{i}|. \tag{4.33}$$

Filling this expression into (4.32) results in

$$|\underline{u}|_{M_2} = |\underline{Z}_a|\,(|\underline{T}_{up}|_M)^2\,\left|\frac{u_X}{u_{M_1}}\right|\,|\underline{i}|,\ \ \text{or} \tag{4.34}$$

$$|\underline{T}_{up}|_M = \sqrt{\frac{|\underline{u}_{M_1}|\,|\underline{u}_{M_2}|}{|\underline{Z}_a|\,|\underline{u}_X|\,|\underline{i}|}}. \tag{4.35}$$

Please note that only electric measurements were necessary to determine the sensitivity coefficient, $|T_{up}|_M$, of the electroacoustic coupler to be calibrated, which does not need to be reversible.

An overall accuracy of 1% can be achieved with this calibration method because electric measurements are very accurate. It is therefore in use in certified calibration laboratories. The conceptual essence of the method is summarized in Fig. 4.20. The point is that by knowing the ratio and the product of two quantities, both of them can be determined.

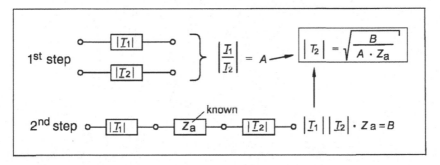

Fig. 4.20. Conceptual summary of the absolute-calibration method

5

Magnetic-Field Transducers

While dealing with magnetic-field transducers in this chapter and electric-field transducers in the next, we will demonstrate that the force-law relationship between the mechanic force, F, and the coupled electric quantity, u or I, is either linear or quadratic. A linear force law, characterized by $F \sim u$ or $F \sim i$, is observed when the energy content of the magnetic or electric field does not vary when the movable component changes position. A quadratic force law, characterized by $F \sim u^2$ or $F \sim i^2$, appears when the movable component meets the electric or magnetic field at a boundary, implying that the energy of the field varies when the movable component changes position. The force in the boundary area is given by the relationships described below that can be derived by imagining a small *virtual shift* of the border area. We have

$$F(x) = -\frac{\mathrm{d}}{\mathrm{d}x} \left[\frac{1}{2} L(x) \, i^2 \right] , \qquad (5.1)$$

for the magnetic field, and

$$F(x) = -\frac{\mathrm{d}}{\mathrm{d}x} \left[\frac{1}{2} C(x) \, u^2 \right] , \qquad (5.2)$$

for the electric field, having recalled from electrodynamics that the energy, $W = \int F(x) \, \mathrm{d}x$, stored in an inductance is $W_{\mathrm{L}} = (L \, i^2)/2$, and the energy stored in a capacitance is $W_{\mathrm{C}} = (C \, u^2)/2$.

Such quadratic force laws must be linearized before they can be used with linear transducers. This linearization is performed by adding a constant offset quantity (bias) to the alternating quantity under consideration. Magnetization bias can, for instance, be achieved with a permanent magnet or a constant magnetization current for magnetic-field transducers, and, similarly, polarization bias can be achieved with an electret or a constant polarization voltage for electric-field transducers.

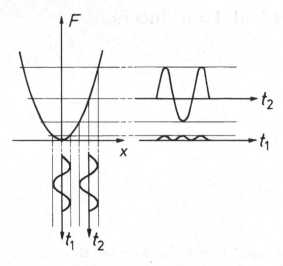

Fig. 5.1. Linearization of force characteristics

Figure 5.1 illustrates the basic idea of this kind of linearization. Adding bias shifts the alternating quantity to a less curved part of the force plot. In mathematical form we get for $x_- \gg x_\sim$,

$$F \sim x^2 = (x_- + x_\sim)^2 = \underbrace{x_-^2}_{1} + \underbrace{2\,x_-\,x_\sim}_{2} + \underbrace{x_\sim^2}_{3}. \tag{5.3}$$

The right side of equation (5.3) has three parts,

- Part 1 denotes a constant quantity. This part can be filtered out with an appropriate high-pass filter

- Part 2 shows a linear relationship with respect to the force. We are primarily interested in the alternating force, $x_\sim$, but it is important to note that its amplitude is controlled by the amount of bias, x_-

- Part 3 becomes more and more irrelevant with increasing bias, that is, for $x_\sim \ll x_-$

Part 3 describes an alternating quantity with a doubled frequency. For for sinusoidal excitation it behaves according to

$$\sin^2 \omega t = \frac{1}{2}\left[1 - \cos(2\,\omega t)\right]. \tag{5.4}$$

Energy-converting elements with an intrinsically quadratic force law do not become transducers in the strict sense until after they have been linearized. It is also only at this point that they become reversible. If they were not linearized, they could only work as sound emitters and not as receivers, and they would only transmit sounds with twice the frequency of the original electric excitation signal.

Magnetic-field transducers are called *velocity transducers* because the output voltage of the internal transducer is proportional to the particle velocity, that is $u_i \sim v_i$. In comparison, electric-field transducers are called *displacement transducers* because the output voltage of the inner transducer is proportional to particle displacement, namely, $u_i \sim \xi_i$. The distinction between velocity and displacement transducers is relevant for optimal mechanic tuning, which will be discussed later.

5.1 The Magnetodynamic Transduction Principle

The Inner Transducer

Consider a rod-shaped conductor exposed to a stationary magnetic field, $\vec{B}$ – drawn in Fig. 5.2. If an electric current, i, passes through this conductor, a force, $\vec{F}$, will act on it according to the expression

$$\vec{F} = i\,(\vec{l} \times \vec{B})\,. \tag{5.5}$$

Moving the rod within the stationary field induces an electric voltage, u, of

$$u = \vec{l}\,(\vec{v} \times \vec{B})\,. \tag{5.6}$$

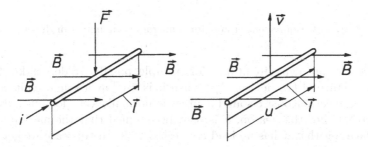

Fig. 5.2. Rod conductor in a stationary magnetic field

The equations are vector equations, but they may be simplified by considering only movements within one spatial dimension. For movements of the rod perpendicular to in the direction of the magnetic induction, $\vec{B}$, the equations for the inner transducer in complex notation reduce to (5.7) and (5.8), The first transducer equation becomes

$$\underline{F}_i = B\,l\,\underline{i}_i, \tag{5.7}$$

and the second one, showing that this is a velocity transducer, becomes

$$\underline{u}_i = B\,l\,\underline{v}_i. \tag{5.8}$$

In these equations, the term $M = B\,l$ is referred to as the *transducer coefficient*. It is also possible to derive (5.8) from (5.7) as shown below, by defining the power at both ports to be equal, which means

$$\underline{u}_i\,\underline{i}_i^* = \underline{F}_i\,\underline{v}_i^*. \tag{5.9}$$

The Real Transducer

Real magnetodynamic transducers[1] have elements besides the inner transducer. A simple equivalent circuit for a real magnetodynamic transducer is given in Fig. 5.3. At the electrical end we see an inductance, L, and a resistance, R. On the mechanical side we have the mass of the movable component, m, the necessary spring, n, and some fluid damping, r.

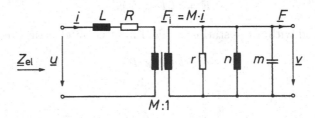

Fig. 5.3. Equivalent circuit for a magnetodynamic transducer

From such an equivalent circuit it is, for example, possible to derive the electric input impedance of the device, $\underline{Z}_{el}$, when it is used as a sound emitter. The efficiency of electroacoustic sound sources is normally very low, usually only a few percent. For this reason, it is usually assumed that the mechanic port is idle, meaning that it has no load connected to it. In this case we get

$$\underline{Z}_{el}\big|_{\underline{F}=0} = j\omega L + R + \left[\frac{M^2}{j\omega\,m + \frac{1}{j\omega\,n} + r}\right]. \tag{5.10}$$

---

[1] The name for this kind of transducer, *magnetodynamic* or just *dynamic*, is of historic origin. In the terminology of mechanics, transducers of all kinds are dynamic devices

5.2 Magnetodynamic Sound Emitters and Receivers

Dynamic Loudspeakers

The dynamic loudspeaker, illustrated in Fig. 5.4, is arguably the most important magnetodynamic sound emitter.

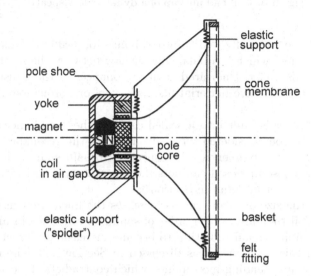

Fig. 5.4. Section view of a dynamic cone loudspeaker

A membrane is elastically supported and driven by a coil carrying alternating
current in a stationary magnetic field. The membrane, usually a cone, plate or
dome, transmits sound into the surrounding air. It should oscillate as a whole,
without any bending waves. Membranes are typically of a *non-deconvolveable*
form, that is, they cannot be unfolded into a plane. They are manufactured
from stiff, sandwich-like layered foils or light foam.

Figure 5.5 illustrates how short coils in a long air gap or long coils in a
short air gap are used to make sure that the coil does not suffer from field
nonlinearities when moving inside the air gap. The long-coil solution allows
for smaller permanent magnets, making it more economical for high magnetic
fields. The resistive part of the coil impedance is normally 4–$8\,\Omega$. The reactive
part, however, can be much larger but may be reduced by a built-in copper
ring. This ring also decreases nonlinearities and increases mechanic damping.
The damping, in turn, decreases the power efficiency of the device.

The power efficiency of a loudspeaker is proportional to the square of the
magnetic-flux density, B, in the air gap. With a common B of 1–2 Tesla, which
is 1–$2\,\mathrm{Vs/m}^2$, the power efficiency is only a few percent. Horn loudspeakers –
as dealt with in Section 8.3 – may achieve up to $15\,\%$.

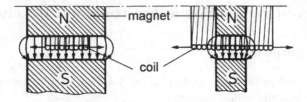

Fig. 5.5. Coil and air gap of a dynamic loudspeaker

Loudspeakers are normally built into cabinets or baffles. This prohibits acoustic shortening, which is a balancing air flow between the front and back sides of the loudspeaker. Undesired cavity resonances of the housing can be dampened with absorptive material like rock wool or porous foam – refer to Section 11.2.

If a loudspeaker is built into a sealed cabinet, the compliance of the enclosed air has to be considered when determining the resonance frequency of the system. At low frequencies, loudspeakers in baffles act as hemispheric radiators while those in closed cabinets (boxes) radiate spherically. We shall analyze this behavior in detail in Section 9.3.

Increasing the size of loudspeakers increases the linear dimensions of the membranes until they are on the order of sound wavelengths in air, causing previously omnidirectional radiation to become increasingly directional. The theory behind this phenomenon is discussed in Section 10.3. The membrane may also enter into bending movements, which contradicts the otherwise increasing directionality. Making the center of the membrane cone a little stiffer encourages the high frequencies to be emitted from a smaller center area, which reduces the effective mass of the system as frequency increases.

Because the cone center moves with the whole membrane, high frequencies are radiated from a moving source, leading to audible *Doppler* shifts. This can be avoided by producing high and low frequencies with different loudspeakers. This arrangement – illustrated in Fig. 5.6 – consists of a tweeter for high frequencies and a woofer for low frequencies. The frequency-cross-over network must be carefully designed.

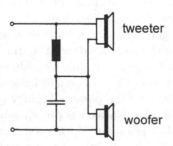

Fig. 5.6. Tweeters and woofers driven through a cross-over network

Magnetodynamic tweeters[2] (dome tweeters) look very much like the moving-coil microphone shown in Fig. 5.10. To avoid focussing directional characteristics at high frequencies, more than one tweeter can be employed. Then, however, a serious problem experienced in all multi-path loudspeaker arrangements has then to be faced an carefully treated, namely, interference of signals radiated from loudspeakers at different locations.

Balancing all the different parameters of loudspeaker systems makes loudspeaker design an art that requires extensive experience and knowledge of materials. A membrane, for example, must not be too stiff, which would cause resonance peaks, but also not too compliant, which would reduce efficiency. The inner membrane support is normally fitted with some damping, the outer support provides an impedance match that prevents the bending waves of the membrane from being reflected.

The mechanic tuning of the loudspeaker usually shifts the main resonance to the low end of the transmitted frequency range. Spherical waves are predominately emitted in this frequency range, and, below a limiting frequency ω', the real part of the radiation impedance, Z_{rad}, increases proportional to ω^2. The radiated power situation depicted in Fig. 5.7 accounts for both of these effects. In Fig. 5.7 it is assumed that the system is excited with a constant force, F, which is equivalent to constant-current excitation.

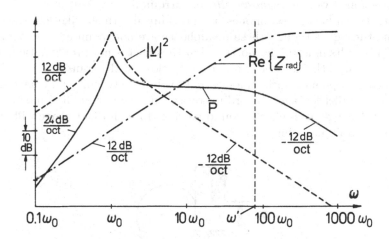

Fig. 5.7. Tuning the frequency response of dynamic loudspeakers

Tuning the mechanic resonance to lower frequencies extends the range. The acceleration, a, of the membrane should be constant to keep the radiated power independent of frequency since $|v^2| \sim 1/\omega^2$ means $|a| = \omega |v| = \mathrm{const}$.

The decrease of power radiation at high frequencies is partly counterbalanced by increased focussing toward receiving points inside the range of the

---

[2] For piezoelectric tweeters see Section 6.4.

radiation beam. Properly designed higher-order resonances further improve the efficiency of the system.

Possible audible *booming*, that is, low-frequency ringing, is avoided by reducing the inner impedance of the electric source until it becomes a matched-power source. This reduces the peak of the resonance. Further reduction of the inner impedance and the current excitation causes the constant-force excitation to convert into a constant-voltage excitation and, thus, into a constant-velocity excitation. Modern power amplifiers can have inner impedances of few mΩ only.

Electric equalization using the techniques discussed here can overcome the mechanic deficiencies of the loudspeaker, but be aware that very large mechanic forces may be needed, reducing the power efficiency. Monitoring the movement of the membrane with sensors and controlling the driving force accordingly is called *motional feedback* and has been successful at reducing distortions – particularly at the lower end of the frequency range.

Dynamic Headphones

There are basically two types of dynamic headphones, open and closed. The open version is traditionally worn *supra-aurally*, or on top of the ear, and the closed version is worn *circum-aurally*, or surrounding the ear.

Open (open-back) headphones are mechanically tuned like loudspeakers (low-end tuning), although these headphones are not mounted in a baffle or box. This results in acoustic shorting, but the effect is not problematic in this instance because the listeners' ears are very close to the transducers.

Closed headphones work on a closed air volume, n_a, which may have some degree of parallel leakage. According to an extremely simplified model that disregards leakage-mass effects, sound passing through this leakage will experience some fluid damping, r_a.

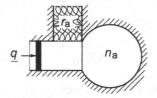

Fig. 5.8. Simplified model of a closed headphone

The situation is depicted in Fig. 5.8. Resulting from

$$\frac{\underline{q}}{\underline{p}} = \mathrm{j}\omega\, n_a + \frac{1}{r_a}, \quad \text{we have} \quad \underline{p} = \frac{\underline{q}}{\mathrm{j}\omega\, n_a + \frac{1}{r_a}}. \tag{5.11}$$

In order to achieve a constant sound pressure, $\underline{p}$, over the entire frequency range, $\underline{q}_0$ and $\underline{v}$ must increase proportionally with ω from the middle frequencies on. This is typically accomplished with high-end tuning and aperiodic damping, where $\underline{\xi}$ is held constant and $v \sim \omega$. At low frequencies, an additional resonance is sometimes applied to create a "fuller" perception of sound.

Dynamic Microphones

Dynamic microphones have been the most widely used consumer microphone type for a long time. Only recently have they been outnumbered by electret and silicon-chip microphones – see Section 6.6.

The *ribbon microphone* type represents the fundamental principle of dynamic microphones in the clearest form. A softly supported ribbon of light conductive material like aluminium moves between the poles of a permanent magnet – illustrated in Fig. 5.9.

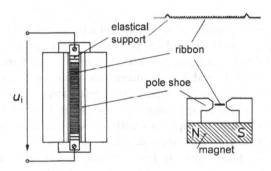

Fig. 5.9. Ribbon microphone

As the sound field reaches the ribbon from both sides, the system acts as a pressure-gradient receiver. It is mechanically low-end tuned. This means that it works in the region above its characteristic frequency, that is, where the ribbon responds as a mass and we have $\underline{v} \sim 1/\omega$. This compensates for sensitivity that would typically decrease with ω. As a direct result of this tuning, the ribbon microphone is unfortunately mechanically delicate and very sensitive to structure-borne sound transmitted from the floor as well mechanic impact. It is usually elastically supported to reduce these effects.

The sensitivity and inner impedance of the ribbon microphone are low since there is only one ribbon in the magnetic field. With a transformer in chain, which that transforms the inner impedance to about $200\,\Omega$, a typical sensitivity would be $T_{\mathrm{up}} \approx 1\,\mathrm{mV/Pa}$.

Ribbon microphones can also be built to form spherical or cardioid receivers by positioning an acoustic sink to one side of the ribbon. Ribbon microphones are rare today but sometimes irreplaceable. Trumpet sounds in studios, for instance, are often picked up with these microphones since they are difficult to overload, meaning that they do not produce nonlinear distortions even at very high sound pressures like $> 130\,\mathrm{dB}$.

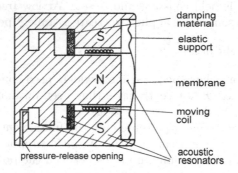

Fig. 5.10. Moving-coil microphone

The *moving-coil microphones* is a type of velocity transducer that increases sensitivity by using a coil instead of a ribbon – depicted in Fig. 5.10. As a result, the induced voltage increases with the number of turns, ν. Unfortunately the mass of the movable component also increases, which may cause audible transient effects due to arrival-time distortions. Typical sensitivities are $T_{\mathrm{up}} > 1.5\,\mathrm{mV/Pa}$. There are many different, sometimes very sophisticated constructions of dynamic microphones with sizes as small as about 20 mm in diameter. Since moving-coil microphones have a low inner impedance, they do not require an amplifier close to the transducer. The coil, however, makes them susceptible to magnetic interference, a susceptibility that can be diminished by putting a fixed compensation coil in series with the moving coil. The mechanic tuning depends on whether the receiver is pressure or pressure gradient and on the directional characteristics.

The input impedance of pressure receivers like the one shown in Fig. 5.10 should be purely resistive to achieve a constant ratio of p/v and a constant resultant output voltage. To accomplish this, the system is usually heavily damped and the main resonance of the system is tuned to middle frequencies – illustrated in Fig. 5.11. Further resonances can be applied at the upper and lower end of the frequency region for the purposes of equalization. These resonances are created by additional cavities behind the membrane in Fig. 5.10, which also shows the felt rings that produce the damping.

Pressure-gradient microphones, that is, figure-of-eight microphones as well as cardioid microphones, can usually be recognized by a second sound inlet on the backside, leading to a carefully specified acoustic delay-line. The system is mechanically tuned to the low end, so that the system acts as a mass in its main operational frequency range. The following construction features can be observed in actual cardioid microphones. Both principles can be employed to decrease the near-field effect.

- The variable-distance principle uses delay lines with different effective lengths for different frequencies in order to increase the driving force at low frequencies

- The double-path principle uses two transducers with different delay lines, one for the high frequencies and one for the low. The two transducers are connected by an appropriate cross-over network

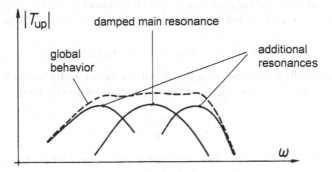

Fig. 5.11. Mechanic tuning of a moving-coil microphone

5.3 The Electromagnetic Transduction Principle

The Inner Transducer

Figure 5.12 illustrates the fundamental arrangement for the inner-transducer principle.

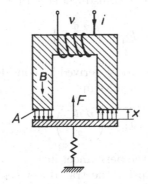

Fig. 5.12. Electromagnet with a movable armature

To derive the transducer equation, we must compute the force on the movable armature. This is done by imagining a small *virtual shift* of the armature, dx.

We start with a fundamental relation from electrodynamics, *Ampere*'s law, which states that in an arrangement as shown in Fig. 5.12, the electric current, i, and the magnetic-flux density, B, are proportional as follows,

$$i\,\nu = (B/\mu_0)\,2x\,, \qquad \nu \ldots \text{ number of turns in the coil} \qquad (5.12)$$

whereby it is assumed that the iron cores of the yoke and the movable armature are highly permeable, so that the energy of the magnetic field is concentrated in the air gap, $2x$. Multiplication with the cross-sectional area of the air gap, A, yields

$$A\,i\,\nu = (A\,B/\mu_0)\,2x = (\Phi/\mu_0)\,2x\,, \qquad (5.13)$$

Φ is the magnetic flux. By inserting the definition of the inductance, $L = \nu\Phi/i$, into (5.13), we get the inductance of our arrangement in the form

$$L = \nu^2\,\frac{\mu_0 A}{2x}\,. \qquad (5.14)$$

Referring to (5.1) and (5.14), we now compute the contracting force as

$$F(x) = -\frac{d}{dx}\left[\frac{1}{2}L(x)\,i^2\right] = \underbrace{\frac{1}{2}\,\frac{i^2\,\nu^2}{2\,x^2}}_{(B/\mu_0)^2}\,\mu_0 A = \frac{B^2 A}{\mu_0} = \frac{\Phi^2}{\mu_0 A}\,. \qquad (5.15)$$

This is clearly a *quadratic* power law. To linearize it, we add a permanent magnetic flux, Φ_-, as a magnetic bias – either by applying a permanent magnet or due to a constant current, i_-. This flux is large in comparison to the alternating flux, $\Phi_\sim$, and the two result in $\Phi = \Phi_- + \Phi_\sim$, which leads to

$$F(x)_\sim \approx \frac{2\,\Phi_-\,\Phi_\sim}{\mu_0 A} = \frac{2\,\Phi_-}{\mu_0 A}\,\frac{\nu\,\mu_0 A}{2x}\,, \quad \text{with } \Phi_\sim = \frac{L\,i_\sim}{\nu} = \nu\,\frac{\mu_0\,A}{2\,x}\,i_\sim\,. \quad (5.16)$$

In this way, we obtain the first transducer equation,

$$\underline{F}_i \approx \left(\frac{\nu\,\Phi_-}{x}\right)\underline{i}_i\,. \qquad (5.17)$$

The second equation can be easily derived by considering the equality of the in- and output power, which results in

$$\underline{u}_i \approx \left(\frac{\nu\,\Phi_-}{x}\right)\underline{v}_i\,. \qquad (5.18)$$

It is important to note that the permanent flux, Φ_-, and the number of turns, ν, appear in the relationship for the transducer coefficient, $M = (\nu\,\Phi_-)/x$. This means that the transducer becomes more efficient and, therefore, more sensitive with increasing magnetic bias and increasing number of coil turns.

The Real Transducer

The equivalent circuit of the real electromagnetic transducer corresponds to the magnetodynamic transducer circuit, with the addition of one very interesting feature. The mechanic compliance is supplemented by a negative compliance called the field compliance, $n_f = \mathrm{d}x/\mathrm{d}F|_{i_-}$, which results from decreasing the air gap[3]. This decreasing gap increases the force attracting the armature and opposing the reversing mechanic force. For large displacements this may cause the membrane to bounce to one of the magnet poles and cling there. This is a very undesired effect that becomes more likely with increasing permanent magnetic bias.

5.4 Electromagnetic Sound Emitters and Receivers

Electromagnetic transducers can be built to be very small and very efficient, but, because of their intrinsically quadratic force law, nonlinear distortions are harder to manage than with magnetodynamic transducers. Examples of traditional applications include telephone-receiver capsules, miniature microphones for hearing aids, pick-up transducers for record players and free-swinging loudspeakers. An electromagnetic telephone-receiver capsule and a hearing-aid microphone are shown in Figs. 5.13 and 5.14 for historical reasons.

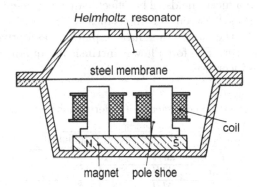

Fig. 5.13. Telephone-receiver capsule

In the traditional telephone capsule, a *Helmholtz* resonator tuned to the middle of the speech spectrum is put on top of the membrane to improve sensitivity to speech signals. Please note that low-frequency tuning would support membrane clinging as explained above.

---

[3] The field compliance can be derived by virtual shift as sketched in the following.
$$\frac{1}{n_f}\bigg|_{i_-} = \frac{\mathrm{d}F}{\mathrm{d}x} = \frac{2\Phi_-}{\mu_0 A}\frac{\mathrm{d}\Phi_-}{\mathrm{d}x} = \frac{2\Phi_-}{\mu_0 A}\frac{-\nu i_-}{2x^2}\mu_0 A = \frac{2\Phi_-}{\mu_0 A}\frac{-\Phi_-}{x} = \left(\frac{\nu\Phi_-}{x}\right)^2\frac{-2x}{\mu_0 A\nu^2} = -\frac{M^2}{L}$$

A special construction trick – shown in Figure 5.14 – has long been used to manufacture efficient miniature microphones. The reed can be made very thin because it is not pre-magnetized. The result is a microphone that is less than 1 cm large and has a sensitivity on the order of $\underline{T}_{up} \approx 1 \, \text{mV/Pa}$.

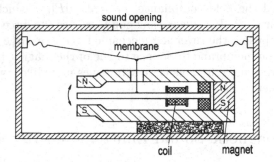

Fig. 5.14. Hearing-aid microphone

5.5 The Magnetostrictive Transduction Principle

Rods made of ferromagnetic material experience a variation of their lengths when exposed to magnetic fields. The effect can be conceptualized by considering the distances between the molecules as forming a *fictive air gap*. This model leads to the same transducer equations as derived for the electromagnetic transducer. The force law is intrinsically quadratic and must be linearized for transducer use.

Table 5.1. Magnetization coefficients of different magnetostrictive materials

Material	Magnetization Coefficient
Iron	$\Delta l/l = -8 \cdot 10^{-6}$
Kobalt	$\Delta l/l = -55 \cdot 10^{-6}$
Nickel	$\Delta l/l = -35 \cdot 10^{-6}$
Ferrite	$\Delta l/l = -100 \text{ to } + 40 \cdot 10^{-6}$

Table 5.1 lists one-dimensional magnetization coefficients, $\Delta l/l$, for a number of magnetostrictive materials. A negative sign means that the rod length decreases when the material is magnetized. Yet, increases do also occur, for example, in ferrit rods. More sophisticated models than the simple air-gap model are obviously necessary to explain this effect.

5.6 Magnetostrictive Sound Transmitters and Receivers

Magnetostrictive transducers have a very high mechanic impedance, which makes them well-suited to underwater and/or ultrasound applications. This type of transducer may achieve efficiencies of more than 90 % when used in water at ultrasound frequencies. Sometimes pre-magnetization is not applied when this principle is used to build narrow-band emitters. In such cases, the transducer emits sound signals with twice the frequency of the exciting electric signals. It is important to remember that in such a context, electric transformers for 60 Hz emit sound at 120 Hz, making them electrostrictive devices.

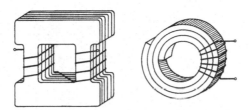

Fig. 5.15. Magnetostrictive transducers

Figure 5.15 shows two exemplary realizations of magnetostrictive transducers. The left one is a two-mass resonator with the bridging bars acting as springs. The different layers of ferromagnetic material are divided by thin isolating foils to avoid eddy currents, which would cause losses. Typical applications include emitters and receivers for echo sounders, and emitters of ultrasound for drilling, cleansing and melding purposes.

6. Magnetostrictive Sound Transmitters and Receivers

Fig. 6.1b.

6

Electric-Field Transducers

In electric-field transducers mechanic forces are caused by electric fields or, in the reverse effect, electric polarization is influenced by mechanic forces. Even more than magnetic-field transducers, electric-field transducers exist in a wide variety of forms and shapes. There are transducers that are intrinsically linear and others that naturally have a quadratic force law and must be linearized. There are also irreversible controlled couplers. In this chapter we concentrate on the basic principles of electric-field transducers and discuss some illustrative examples.

6.1 The Piezoelectric Transduction Principle

Certain crystals may be exploited as transducers because they have the following properties. (a) The crystal's physical dimensions change when an electric field is applied to it. (b) Deformations caused by mechanic forces cause electric polarization on the surfaces of the crystal. These effects are subsumed under the term *piezoelectricity* (pressure electricity). In order for piezoelectric effects to occur, it is essential that the crystals have no center of symmetry – illustrated in Fig. 6.1 (b).

Two characteristic material parameters are used to characterize piezoelectricity, the piezoelectric coefficient, e, and the piezoelectric module, d. These parameters are defined by the following equations,

$$\sigma = e\,s, \quad \text{and} \quad \sigma = d\,\theta, \tag{6.1}$$

where σ is the electric polarization, $s = \xi/x$ is the strain (amount of stretching) and θ is the stress (mechanic tension). The piezoelectric module, d, is

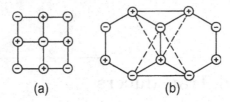

Fig. 6.1. Crystal lattices, (a) with, (b) without center of symmetry

often preferred over e by application experts[1], but in this chapter we shall work with e for systematic reasons.

There are generally six different strains that may exist in solid bodies. Three of these exist as normal strains in the directions of the spatial coordinates, x, y, z, and the other three are shear strains along the planes, xy, yz, and xz. As an example, the dielectric flux density, D, for just the x-direction is expressed as

$$D_x = \sum_i e_{x,i}\, s_i + \sum_j \varepsilon_{x,j}\, E_j, \qquad \varepsilon \ldots \text{dielectric permittivity} \qquad (6.2)$$

with $i \ldots x$, y, z, xy, yz, xz, and $j \ldots x$, y, z. In many materials, however, most of the piezoelectric coefficients are zero. We shall only deal with one coefficient at a time in the following examples. Due to linearity, the total result can be obtained by superposition.

Important materials with *inherent piezoelectricity* are the natural crystals *quartz* and *turmaline*, and the artificial crystals *potassium sodium tartrate* (known as *Rochelle* or *Seignette* salt), *lithium niobate, lithium sulfate hydrate* and *cadmium sulfide*. Table 6.1 lists exemplary material characteristics, where ε_0 is the permittivity of the vacuum. Quartz is particularly stable with regard to temperature and has very low internal losses. *Rochelle* salt has a high piezoelectric effect but is very sensitive to temperature and humidity.

In addition to the materials noted above, many electrostrictive substances can be treated to become piezoelectric. We shall deals with this *influenced piezoelectricty* in Section 6.3. .

The Inner transducer

Now we will derive the piezoelectric transducer equation for the longitudinal piezoelectric effect, which is incorporated in the quartz thickness vibrator shown in Fig. 6.2. The other five dimensions would be derived similarly.

As stated earlier, we have $\sigma = e\, s$ and $\xi = s\, x$, from which we get $\sigma = e\, \xi / x$. This allows us to write

---

[1] It is possible to derive d from e with the modulus of elasticity, *Young*'s modulus, being known and linearity according to *Hook*'s law being assumed

Table 6.1. Typical characteristics of inherently piezoelectric materials*)

Name	Material	Piezoelectric coefficient, e	Piezoelectric module, d	Dielectric permittivity, ε	Remark
Quartz, SiO$_2$	Natural crystal	170 mC/m^2 (normal strain)	2.3 pC/N	4.6 ε_0	*Curie* temp. 570 °C
Rochelle salt	Synthetic crystal	4.7 C/m^2 (shear strain)	2300 pC/N	200 ε_0 to 1300 ε_0	*Curie* temp. 24–35 °C

*) [C/N] = [m/V], $\varepsilon_0 = 8.854188$ pF/m,

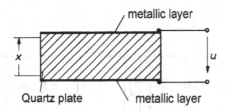

Fig. 6.2. Arrangement for the longitudinal piezoelectric effect

$$A\,\sigma = Q_{\mathrm{el}} = A\frac{e}{x}\xi\,, \tag{6.3}$$

where $Q_{\mathrm{el}} = \sigma A$ is the electric charge. By making use of $i(t) = \mathrm{d}Q_{\mathrm{el}}/\mathrm{d}t$ and $v(t) = \mathrm{d}\xi/\mathrm{d}t$, and by assuming that the power is equal at the two ports, that is $\underline{F}\,\underline{v}^* = \underline{u}\,\underline{i}^*$, we arrive at the two transducer equations,

$$\underline{F} = \left(\frac{e\,A}{x}\right)\underline{u}, \quad \text{and} \quad i = \left(\frac{e\,A}{x}\right)v\,. \tag{6.4}$$

The transducer coefficient, N, thus, comes out as

$$N = \frac{e\,A}{x}\,. \tag{6.5}$$

Piezoelectric transducers, like all electric-field transducers, are elongation (displacement) transducers. With the capacitance being $C = A\varepsilon/x$, we get

$$\underline{Q}_{\mathrm{el}} = A\frac{e}{x}\xi = C\,\underline{u}\,, \tag{6.6}$$

which results in

$$\underline{u} = \frac{e}{\varepsilon}\,\underline{\xi}\,, \quad \text{that is} \quad \underline{u} \sim \underline{\xi}\,. \tag{6.7}$$

Note that the ratio of e/ε is the relevant quantity for characterizing the distinction of the piezoelectric effect.

As stated above, we consider the inner transducers to be lossless. Yet, in ferroelectric materials there may be losses due to hysteresis. These can be accounted for in the equivalent circuit of the real transducer as shown below.

The Real Transducer

There is a gyrator in the equivalent circuit based on electromechanic analogy #2. To avoid this, the mechanic part of the circuit can be replaced with its dual, which is actually equivalent to switching to analogy #1. Figure 6.3 shows a simplified equivalent circuit in both forms. The bottom one is usually found in the literature.

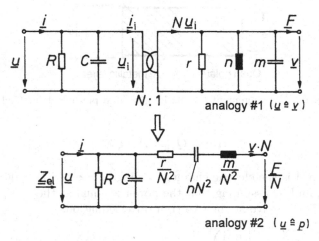

Fig. 6.3. Equivalent circuits for electric-field transducers

The electric input impedance is most easily derived from analogy #1 and is equal to

$$\underline{Z}_{el}\Big|_{\underline{F}=0} = \frac{1}{G + j\omega C + \left[\frac{N^2}{r + j\omega\, m + \frac{1}{j\omega\, n}}\right]}. \tag{6.8}$$

The equivalent circuits depicted above can also be applied to the other kinds of electric-field transducers.

6.2 Piezoelectric Sound Emitters and Receivers

Piezoelectric sound emitters and receivers have a wide range of applications involving airborne, waterborne and solid-borne sound. Their realizations depend heavily on the particular application. Figure 6.4 provides an overview of more important vibration forms.

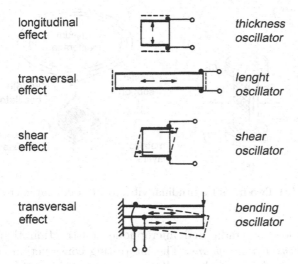

Fig. 6.4. Various vibration forms induced by piezoelectricity

The bending oscillator is especially interesting. Two piezoelectric layers are glued back-to-back with a conducting foil in between – shown in Fig. 6.5. High capacitance, C, results in a low inner electric impedance that allows for long connecting wires. The mechanic impedance is also relatively low, which provides for an adequate impedance match, particularly in airborne-sound fields. Since the two layers act electrically in parallel, good electrical shielding is also achieved.

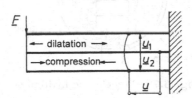

Fig. 6.5. Piezoelectric layers as a bending vibrator

A traditional domain for piezoelectric sound emitters is underwater applications, especially utilizing ultrasound. Figure 6.6 (a) shows a two-mass longitudinal vibrator that finds one of its applications in echo sounders. The transducer is reversible, allowing it to also be used as an underwater sound receiver, also called a *hydrophone*.

Piezoelectric sound receivers exist in a variety of forms. A relatively high capacitance of about 1–3 µF is valuable from a practical standpoint because it allows for a few meters of cable before amplification becomes necessary.

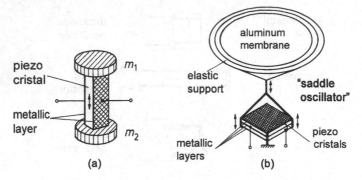

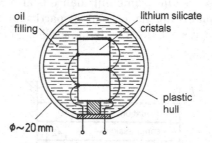

Fig. 6.6. (a) Two-mass longitudinal vibrator, (b) crystal microphone

Figure 6.6 (b) shows a traditional microphone construction that has been known as the *crystal microphone*. The interesting construction trick, worth mentioning here, is that a saddle oscillator is used to achieve low compliance.

   This kind of microphone has high-frequency tuning when used as a pressure receiver and mid-frequency tuning with high damping when used as a pressure-gradient receiver. To understand these rules, recall that they are elongation receivers, and $\underline{\xi} = \underline{v}/j\omega$.

Fig. 6.7. Piezoelectric hydrophone

Figure 6.7 shows a hydrophone where several piezoelements are mechanically stacked (cascaded). Stacking is a widely used method to increase the sensitivity of piezoelectric devices. Piezoelectric transducers are also used as acceleration and/or velocity sensors in vibration meters. Three different construction principles are depicted in Fig. 6.8.

   Piezoelectric resonators are used in large quantities as frequency-determining elements in electric circuits, for example, quartz filters in electric watches. Quartz is a preferred material here because of its low internal damping and high thermal stability. The quartz crystal is operated in a vacuum in order to achieve very high quality factors on the order of $Q \approx 5 \cdot 10^5$.

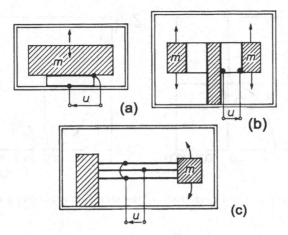

Fig. 6.8. Acceleration/velocity sensors

Figure 6.9 illustrates an equivalent circuit for a quartz filter. There are two resonances, one parallel and one serial, that lie very closely together, so that $\omega_p = 1.01\,\omega_s$. The formulae for the resonance frequencies are

$$\omega_s = \frac{1}{\sqrt{L_m\,C_m}}, \quad \text{and} \quad \omega_p = \frac{1}{\sqrt{L_m(\frac{C\,C_m}{C+C_m})}}. \tag{6.9}$$

The curve of the impedance and admittances of the quartz filter over frequency can roughly be estimated by applying *Foster*'s reactance rules, which are well known to network specialists. Figure 6.10 shows schematically the resulting plots of the reactance and the magnitude of the input-impedance as a function of frequency[2].

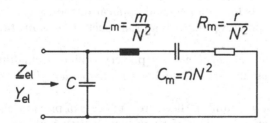

Fig. 6.9. Equivalent circuit for a quartz filter

---

[2] When plotting such curves, it is often advantageous to start with the lossless case and then introduce small losses later

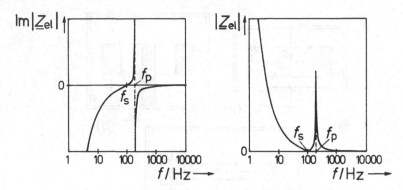

Fig. 6.10. Reactance and magnitude of the input impedance of a quartz filter

6.3 The Electrostrictive Transduction Principle

All dielectric materials experience mechanic deformations when exposed to an electric field. There is a basic quadratic relationship between the strength of the field and the stress of the material, which can, however, be linearized with the bias of a permanent electric field. It is in this way that the electrostrictive effect becomes reversible and can be exploited for transducer use.

A particularly strong electrostrictive effect can be observed in ferroelectric materials. Some of these materials can be permanently polarized by heating them up above their *Curie* temperature and letting them cool down while being exposed to a strong electric field. The ferroelectric *Weiss* domains orient themselves according to the direction of the electric field and keep this orientation when cooled down.

The behavior of polarized electrostrictive materials is called *influenced piezoelectricity*. It acts equivalently to inherent piezoelectricity. Literature often does not even distinguish between inherent and influenced piezoelectricity. Important materials for technological applications are,

(a) Ceramic (polycrystaline) materials such as *barium titanate* and *lead zircone titanate*,

(b) Amorphous *piezopolymeres* (polycrystalline or high-polymere semi-crystaline artificial materials), such as *polyvinyliden fluoride* and *polyvinyl cloride*.

Among the piezoceramics, there are a number of proven compositions with different characteristics, some of them having very high piezoelectric coefficients. Piezopolymers obtain their piezolectric features by being stretched in one direction first and subsequently being polarized permanently. Transducer equations and simple equivalent circuits are identical for inherent and influenced piezoelectricity.

Some example material characteristics are compiled in Table 6.2. Note that the material in the third line is not piezoelectric but an electret. Section 6.6 will give more details on electrets.

Table 6.2. Characteristics of materials with influenced piezoelectricity (line 1 and 3). The material in line 3, shown for comparison, is a piezoelectret[*)]

Name	Material	Piezolelectric coefficient, e	Piezoelectric module, d	Dielectric permittivity, ε	Remark
Lead-zircone titanat, PZT	Ceramic	$\approx 20\,\mathrm{C/m^2}$ (shear strain)	$300\,\mathrm{pC/N}$	$1400\,\varepsilon_0$ to $1700\,\varepsilon_0$	*Curie* temp. $\approx 350\,^{\circ}\mathrm{C}$
Polyvinyliden fluoride, PVF$_2$	Polymer	$50\,\mathrm{mC/m^2}$ (transversal strain)	$25\,\mathrm{pC/N}$	$13\,\varepsilon_0$	*Curie* temp. $> 100\,^{\circ}\mathrm{C}$
Polypropylene cellular film, PP	Polymer	$0.50\,\mathrm{mC/m^2}$ (normal str.)	$500\,\mathrm{pC/N}$	$1.5\,\varepsilon_0$	temp. limit $\approx 50\,^{\circ}\mathrm{C}$

[*)]$[\mathrm{C/N}] = [\mathrm{m/V}]$, $\varepsilon_0 = 8.854188\,\mathrm{pF/m}$

6.4 Electrostrictive Sound Emitters and Receivers

Among the electrostrictive sound emitters and receivers there are many shapes in addition to those used in piezoelectric transducers. It is indeed the great advantage of influenced piezoelectricity that many relevant materials are freely formable. Piezopolymeres can even be manufactured into foils (films) as thin as a few μm that are elastic and can be stretched across curved surfaces. Figures 6.11 and 6.12 present a collection of realized forms.

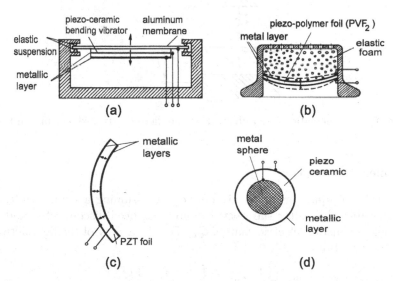

Fig. 6.11. (a) Piezoceramic telephone receiver, (b) headphone with piezopolymer foil, (c) ultrasound beamer with piezopolymer foil, (d) piezoceramic hydrophone for ultrasound (diameter $\leq 0.5\,\mathrm{mm}$)

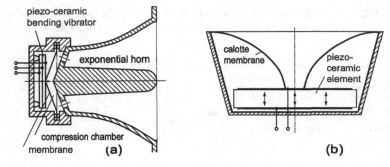

Fig. 6.12. (a) Piezoceramic horn tweeter, (b) piezoceramic calotte tweeter. NB: Due to their high impedance, piezoelectric tweeters may be operated in parallel with dynamic loudspeakers without cross-over networks

6.5 The Dielectric Transduction Principle

Historically this principle was called the *electrostatic* principle. The result of the arrangement shown in Fig. 6.13 is a quadratic force law. Two conducting plates, one fixed and one movable, are exposed to an electric voltage, u. The plates become electrically charged, resulting in an electrostatic pulling force.

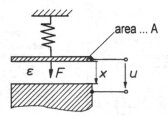

Fig. 6.13. Schematic sketch illustrating the dielectric transducer principle

The Inner Transducer

The transducer equations are again derived by assuming a virtual shift, dx, of the movable plate after the electric source has been disconnected and the plates hold a constant electric charge, Q_{el}. For the energy of the field between the plates, we obtain

$$W_C = \frac{1}{2} C u^2 = \frac{1}{2} \frac{\varepsilon A}{x} u^2 = \frac{1}{2} \frac{Q_{el}^2}{C}, \tag{6.10}$$

with $Q_{el} = C u$ and $C = \varepsilon A/x$. With the electric charge, Q_{el}, held constant, a virtual shift of dx reveals a quadratic power law as follows.

$$F(x) = -\frac{d}{dx}\left[\frac{1}{2}C(x)u^2\right] = \frac{Q_{el}^2}{2\varepsilon A} = \frac{C^2}{2\varepsilon A}u^2 = \frac{A\varepsilon}{2x^2}u^2 = \frac{1}{2}\frac{C}{x}u^2. \quad (6.11)$$

After linearizing with a DC voltage, so that $u = u_- + u_\sim$ with $u_- \gg u_\sim$, and assuming equality of the port powers, that is $\underline{F}\,\underline{v}^* = \underline{u}\,\underline{i}^*$, the following transducer equations result,

$$\underline{F} \approx \left(\frac{C\,u_-}{x}\right)\underline{u}, \quad \text{and} \quad \underline{i} \approx \left(\frac{C\,u_-}{x}\right)\underline{v}. \quad (6.12)$$

The dielectric transducer coefficient then becomes

$$N = \left(\frac{C\,u_-}{x}\right). \quad (6.13)$$

The dielectric transducer is an elongation (displacement) transducer since, with $u = Q_{el}/C$, it follows that $du \sim dx$ and $\underline{u} \sim \underline{\xi}$.

The Real Transducer

The equivalent circuit for the real dielectric, linearized transducers is identical to the one shown in Fig. 6.3 for piezoelectric transducers. Electrical losses, however, can be neglected in praxi. The mass of the membrane and, at least for pressure receivers, the mechanic damping are made very small compared to the compliance of the air cushion. In other words, the electric input impedance becomes practically a capacitance, which results in

$$\left.\underline{Z}_{el}\right|_{\underline{F}=0} \approx \frac{1}{j\omega(C + n\,N^2)}. \quad (6.14)$$

As with electromagnetic transducers, a detailed analysis shows that a negative field compliance appears in parallel with the compliance of the air volume. As a result, we again face some danger of the membrane jumping to one plate and clinging to it, especially when the polarization voltage is too high.

6.6 Dielectric Sound Emitters and Receivers

Dielectric sound emitters are used as loudspeakers, especially tweeters, and as ultrasound emitters – see Fig. 6.14. When light, tightly stretched membranes are used, mechanic tuning to the high-frequency end is unavoidable but may be compensated for with equalization in the electric circuit.

Since the membrane experiences only very small displacements, large areas that move in phase are necessary for efficient sound emission. The efficiency increases proportional to the square of the polarization voltage, but high voltages carry the danger of electric burn-throughs. Fortunately, there are self-healing membrane materials. A variety of shapes are possible, including large

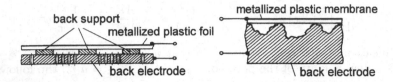

Fig. 6.14. Examples of dielectric sound emitter constructions

planes and spheres, whereby wire meshes may be used for the back electrodes. Loudspeakers have been built with a frequency range down to 50 Hz using this principle. Figure 6.15 shows two realized push-pull arrangement. There are also dielectric (electrostatic) headphones, which are known for their particularly "clear" presentation. This auditory clearness, which also holds for electrostatic loudspeakers, can be attributed to the small moving masses that do not cause any severe phase distortions.

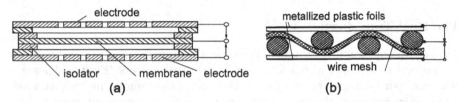

Fig. 6.15. Electrostatic loudspeaker with push/pull driving forces, **(a)** arrangement as used in high-quality loudspeaker, **(b)** arrangement as, e.g., used in spherical loudpeakers

Dielectric sound receivers are known as *condenser microphones*. They have a broad frequency range and only minimal distortions, making a high-quality condenser microphone a good choice for studio and measuring applications.

Figure 6.16 presents a schematic section of such a microphone. The air gap behind the membrane is typically 5–100 μm, and capsule capacitances are 20–100 pF for studio and measuring microphones. The membranes are usually gold-coated plastic foils. Polarization voltages are 40–200 V. The losses of the capsule capacitance are very low, represented by a so-called charge resistance as high as 0.5–300 GΩ. Due to the high inner impedance, the connecting wires are prone to induce noise, such as humming. Further, they add a parasitic capacitance and may cause mechanical problems. To avoid this, there is often an impedance-converting amplifier positioned back-to-back with the capsule. After this stage, typical sensitivities are about $T_{up} \approx 10\text{–}30\,\text{mV/Pa}$.

Condenser microphones, when driven with in a so-called *low-frequency circuit* – shown in Fig. 6.16 – have, on principle, a low-end roll-off that is determined by the 1st-order R/C high-pass formed by the capsule capacitance and the charge resistance.

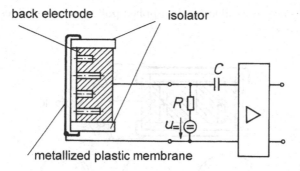

back electrode   isolator

metallized plastic membrane

Fig. 6.16. Condenser microphone with accompanying electric circuit

An alternative method of operating dielectric microphones, the so-called *high-frequency circuit*, avoids this high-pass behavior. Here, as a result of their varying capacitance, condenser microphones are integrated into electric-oscillator circuits as frequency-determining elements. However, in this mode, the condenser microphones no longer act as reversible transducers but as controlled couplers. The principle is applied for measurement microphones because it is extremely resistant to induced noise. Also, as it does not have a lower cut-off frequency, you can actually measure the static air pressure with it!

Condenser microphones are elongation transducers. As pressure receivers they are high-end tuned, and as figure-of-eight or cardioid microphones – such as shown in Fig. 6.17 – they are low-end tuned with prominent damping.

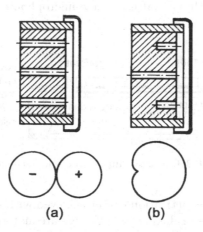

(a)   **(b)**

Fig. 6.17. Figure-of-eight or cardioid microphone

High symmetry of the figure-of-eight directional characteristics is achieved by placing a counter electrode on the other side of the capsule. This also allows the directional characteristics to be steered electrically since the sensitivities of the membranes are controlled by the applied polarization voltages. Figure 6.18 provides an overview.

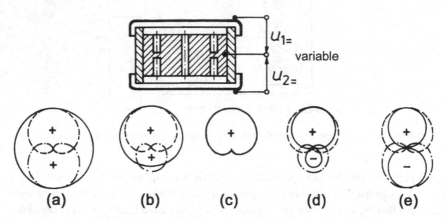

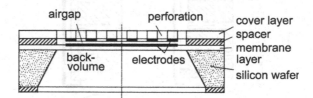

Fig. 6.18. Microphone with steerable directional characteristic, (**a**) sphere (**b**) wide-angel cardiod, (**c**) cardiod (**d**) super cardiod, (**e**) figure-of-eight

Miniature dielectric microphones are micromanufactured very much like electronic chips are produced, for instance, by etching little membranes with an air gap into the substrate of silicon wafers. This leads to silicon-chip microphones – schematically shown in Fig. 6.19. Due to their very small dimensions, chip microphones can easily be arranged to form arrays. Both during the production process and the actual use, these microphones can withstand fairly high temperatures.

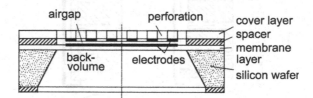

Fig. 6.19. Silicon-chip microphone (schematic)

In a similar way as magnetization can be achieved with magnets, polarization can be achieved with so-called *electrets*. These are materials that show permanent polarization after appropriate treatment, such as exposure of heated material to strong electric fields and subsequent down-cooling, corona discharge or other forms of electron bombardment. If such a material, for example *teflon* or *flourcarbon*, is positioned behind the membrane, an external polarization voltage becomes superfluous. Actually, these electrets easily mimic polarization voltages of 100 V.

Figure 6.20 illustrates the principle. Electret microphones, including an integrated impedance converter, can be manufactured very economically and have become a widespread microphone type worldwide.

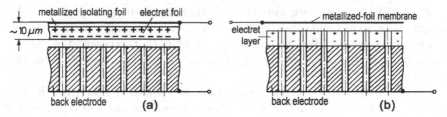

Fig. 6.20. Electret microphone, (a) version with electret membrane, (b) version with electret back electrode

Recently, *elastic electrets* have become available. An example would be piezo-electric polymere films, made from *polyethylene*. They are extruded and treated mechanically afterwards in such a way that they develop a cellular structure with ample void bubbles in it – sketched in Fig. 6.21.

Polarization is performed by corona discharge with the effect that surface polarization develops around the voids. These so-called *piezoelectrets* behave very much like piezoelectric material – see Table 6.1 for comparison. If piezoelectret films are coated with conducting layers, they act as microphones without further attachments. With stacked $\approx$ 50-μm layers of these films, sensitivities of $\geq 20\,\mathrm{mV/Pa}$ can be achieved.

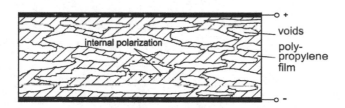

Fig. 6.21. Piezoelectret film

6.7 Further Transducer and Coupler Principles

In addition to the transducer principles that we already dealt with in this and the prior sections, there are some further techniques, even without magnetic and electric fields that we would like to briefly mention.

The first is based on the fact that a wire with a varying current passing through it heats up, causes variations of the air pressure around it, and radiates sound. This is called a *thermophone*. The resistance of the wire will also vary in response to the varying velocity of air flow, making it useful as a *velocity receiver*. This is called a *hot-wire anemometer*. Modern constructions employ micromachined sensors with two very thin, heated wires. These cover a frequency range from 0 Hz up to more than 20 kHz.

A second transducer exploits the fact that a high-frequency glowing discharge emits sound when modulated with an audio signal, creating the basis for an *ionophone*. The reverse effect, namely, that exposing the glowing discharge to airborne sound changes its resistance in synchrony, is called a *cathodophone*.

Further methods for picking up sound or evaluating vibrations include the *laser interferometer* and *optical microphones*. The latter is comprised of a light-conducting glass fiber that modulates light when deflected by sound. These microphones can withstand high temperatures, making them useful, for example, in the measurement of sound inside combustion engines. Further, silicon-chip microphones have been built in such a way that the membrane movements are monitored by optic rather than electric means. Optical microphones are insensitive to electromagnetic fields, which can be an advantage in adverse environments. They are not reversible, which means that they are controlled couplers and not transducers.

7

The Wave Equation in Fluids

So far in this book we have dealt with vibrations. These are processes that vary as functions of time. We were able to describe relevant types of vibrations with *common* differential equations. This chapter now focuses on waves, which are processes that vary with both time and space. Their mathematical description requires *partial* differential equations.

Motivated by the electroacoustic analogies, we start this chapter with an excursion into electromagnetic waves. This excursion will bring us back to sound waves in the next section[1].

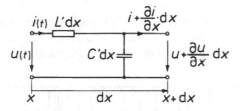

Fig. 7.1. Equivalent circuit for a differential section of a homogeneous, lossless electrical transmission line

Figure 7.1 illustrates the equivalent circuit for a elementary section of a homogeneous, lossless electrical transmission line. For this section, the following loop and node equations hold,

---

[1] Readers who have not yet heard of electromagnetic waves may wish to read Sections 7.1–7.2 first

$$u - \left(u + \frac{\partial u}{\partial x}\,\mathrm{d}x\right) = L'\,\mathrm{d}x\,\frac{\partial i}{\partial t}, \quad \text{and} \tag{7.1}$$

$$i - \left(i + \frac{\partial i}{\partial x}\,\mathrm{d}x\right) = C'\,\mathrm{d}x\,\frac{\partial u}{\partial t}. \tag{7.2}$$

The following two linear differential equations are obtained by neglecting higher-order differentials,

$$-\frac{\partial u}{\partial x} = L'\,\frac{\partial i}{\partial t} \quad \text{and} \quad -\frac{\partial i}{\partial x} = C'\,\frac{\partial u}{\partial t}. \tag{7.3}$$

This set of equations shows that a temporal variation of the slope of one of the variables results in a proportional spatial variation of the slope of the other. This causes that the total energy on the line swings between two types of complementary energies, namely,

- magnetic energy per length, $W' = \frac{1}{2}L'\,i^2$
- electric energy per length, $W' = \frac{1}{2}C'\,u^2$

The two linear differential equations (7.3) can be combined into a differential equation of second order, which is

$$\frac{\partial^2 u}{\partial x^2} = \frac{1}{c_{\text{line, el}}^2}\,\frac{\partial^2 u}{\partial t^2}. \tag{7.4}$$

This is the so-called *wave equation,* here in the formulation for electromagnetic waves on an electrical transmission line. Hereby $c_{\text{line, el}}$ is the propagation speed of electromagnetic waves on the transmission line, namely,

$$c_{\text{line, el}} = \frac{1}{\sqrt{L'\,C'}}. \tag{7.5}$$

In acoustics we also have waves that propagate along one coordinate, for example, in gas-filled tubes with small diameters compared to wavelength. Such *longitudinal* compression waves are schematically sketched in Fig. 7.2.

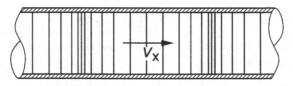

Fig. 7.2. One-dimensional longitudinal acoustic wave. Zones of compression and rarefaction are schematically indicated

Two complementary forms of energies are as well required for these types of waves to occur, but in this case we choose energies per volume to be compatible with the use of p and v as characteristic sound-field quantities. We thus get

- potential energy per volume, $W'' = \frac{1}{2}\kappa p^2$ κ ...volume compressibility
- kinetic energy per volume, $W'' = \frac{1}{2}\varrho v^2$ ϱ ...mass density

Mimicking the wave equation for the electromagnetic waves, we now suggest the following wave equations for one-dimensional acoustic waves,

$$-\frac{\partial p}{\partial x} = \varrho\,\frac{\partial v}{\partial t} \quad \text{and} \quad -\frac{\partial v}{\partial x} = \kappa\,\frac{\partial p}{\partial t}\,, \tag{7.6}$$

and the combined second order expression as

$$\frac{\partial^2 p}{\partial x^2} = \frac{1}{c^2}\frac{\partial^2 p}{\partial t^2}\,, \tag{7.7}$$

with the propagation speed of sound waves to be

$$c = \frac{1}{\sqrt{\kappa\,\varrho}}\,. \tag{7.8}$$

The pressing question now is whether these supposed equations, which are so neatly analogous to electric expressions, are actually in compliance with physical reality. The answer is *yes*, at least approximately. Yet, there are a number of features of the media where sound exists that must be idealized. The following section will elaborate on them.

7.1 Derivation of the One-Dimensional Wave Equation

We assume an idealized medium as a model of the real physical medium. Only compression and expansion but no shear stress are allowed in this medium, which limits us to a category of media called *fluids*. Many gases and liquids can be treated as fluids, which only experience longitudinal waves. The following features of the idealized medium are assumed.

- The medium is homogeneous and does not crack, meaning that there are no inclusions of vacuum as might be caused by cavitation

- The thermal conductivity is zero, which assumes adiabatic compression

- The inner friction is zero, meaning that there are no energy losses and no viscosity

- The medium has defined mass and elasticity

We now assume that the medium is not flowing and that there is no drift, meaning that $v_- \approx 0$. The alternating part of the pressure be small compared to the static pressure, that is, $p_- \gg p_\sim$. The alternating part of the density is also small then in comparison to the static one, $\varrho_- \gg \varrho_\sim$. This results in a so-called *small-signal operation* of the medium.

We must recollect three fundamental physical relationships in order to arrive at the wave equation. These are listed in the following.

- *Hook*'s law, applied to fluids
- *Newton*'s mass law
- The mass-preservation law

These three relationships will now be formulated in their differential forms that are required for the rest of our discussion. In differential form, they are known as

- State equation
- *Euler*'s equation
- Continuity equation

The State Equation

The relationship $p = f(\varrho)$ applies to the medium because it has mass and elasticity. For small-signal operation, that is $\varrho_\sim \ll \varrho_-$, we substitute this function with its tangent at the operating point, that is

$$p_\sim = \left.\frac{\partial p}{\partial \varrho}\right|_{\varrho_-} \varrho_\sim \quad \text{or} \quad p_\sim = c^2 \varrho_\sim . \tag{7.9}$$

This leads us to the *state equation* in the form

$$\partial \varrho = \frac{1}{c^2} \left.\partial p\right|_{\varrho_-} \quad \text{or} \quad c = \sqrt{\left.\frac{\partial p}{\partial \varrho}\right|_{\varrho_-}} . \tag{7.10}$$

In this way we make our *first linearization* by assuming a linear spring characteristic for the fluid. We shall later prove that c is actually the speed of sound, which is specific for a material but are also dependent on its temperature and static pressure.

The proportionality coefficient in the state equation, c^2, can be estimated theoretically for most single-atom gases by assuming a *perfect gas*. If such a gas is compressed in the absence of heat conduction, we have so-called *adiabatic compression*. The then applicable adiabatic law of thermodynamics states that

$$p\, V^\eta = \text{const} = p_-\, V_-^\eta , \tag{7.11}$$

where V is the volume of a mass element concerned and $\eta = c_p/c_v$ (usually known as γ) is the ratio of the specific heat capacities[2], with which we derive

---

[2] These quantities are taken from thermodynamics, where c_p is the specific heat capacity at constant pressure and c_v the specific heat capacity at constant volume

$$p = \left(\frac{V_-}{V}\right)^{\eta} p_- = \left(\frac{\varrho}{\varrho_-}\right)^{\eta} p_- . \tag{7.12}$$

Differentiation of this expression at the position $\varrho = \varrho_-$ renders the following,

$$\left.\frac{\partial p}{\partial \varrho}\right|_{\varrho_-} = \eta \underbrace{\left(\frac{\varrho}{\varrho_-}\right)^{\eta-1}}_{1 \text{ for } \varrho = \varrho_-} \frac{1}{\varrho_-} p_- , \tag{7.13}$$

and then

$$\left.\frac{\partial p}{\partial \varrho}\right|_{\varrho_-} = c^2 = \frac{\eta\, p_-}{\varrho_-} . \tag{7.14}$$

Finally we get for the speed of sound in the perfect gas,

$$c = \sqrt{\frac{\eta\, p_-}{\varrho_-}} = \sqrt{\frac{1}{\varrho_-\, \kappa_-}} \quad \text{whereby} \quad \kappa_- = \kappa\, |_{\varrho_-} = 1/(\eta\, p_-) . \tag{7.15}$$

The expression is written in a similar way for liquids, but there the reciprocal value of the volume compressibility, the compression module $K = 1/\kappa$, is more often used. Some specific material values are given in Table 7.1.[3]

Table 7.1. Values of typical materials

Material	Density, ϱ_-	Sound speed, c	Characteristic impedance, Z_w
Air at 1000 hPa and 20°C	$1.2\,\mathrm{kg/m^3}$	$343\,\mathrm{m/s}$	$412\,\mathrm{Ns/m^3}$
Water at 10°C	$1000\,\mathrm{kg/m^3}$	$1440\,\mathrm{m/s}$	$1.44 \cdot 10^6\,\mathrm{Ns/m^3}$
Steel (longitudinal wave)	$\approx 7500\,\mathrm{kg/m^3}$	$\approx 6000\,\mathrm{m/s}$	$\approx 45 \cdot 10^6\,\mathrm{Ns/m^3}$

Euler's Equation

Figure 7.3 depicts a differential mass of gas in a tube with rigid walls and a diameter small compared to the wavelength. The mass element is accelerated due to a pressure difference, and the total differential of the particle velocity, v, a function of time and space, is

$$\mathrm{d}v\,(x, t) = \frac{\partial v}{\partial t}\mathrm{d}t + \frac{\partial v}{\partial x}\mathrm{d}x . \tag{7.16}$$

From this we arrive at the acceleration,

$$a = \frac{\mathrm{d}v}{\mathrm{d}t} = \frac{\partial v}{\partial t} + \frac{\partial v}{\partial x}\frac{\mathrm{d}x}{\mathrm{d}t} , \quad \text{where} \quad \frac{\mathrm{d}x}{\mathrm{d}t} = v . \tag{7.17}$$

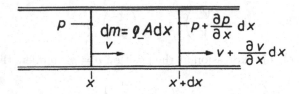

Fig. 7.3. An accelerated differential mass, $\mathrm{d}m$, of fluid in a tube

According to *Newton's* mass law, $F = m\,a$, we can now write

$$A\left[p - \left(p + \frac{\partial p}{\partial x}\mathrm{d}x\right)\right] = (\varrho\,A\,\mathrm{d}x)\left(\frac{\partial v}{\partial t} + \frac{\partial v}{\partial x}v\right), \tag{7.18}$$

which is *Euler's equation*.

Please note that a *second linearization* is implemented by replacement of ϱ with its static value, ϱ_-. Further, the second term in the sum, $v\,\partial v/\partial x$, is usually irrelevant in acoustics since v is zero in resting media, and also because the variation of v over space is usually small, which means that the term is actually a differential of the second order. Neglecting it is our *third linearization*[4]. We thus end at our first linear wave equation as follows,

$$-\frac{\partial p}{\partial x} = \varrho_-\frac{\partial v}{\partial t}. \tag{7.19}$$

The Continuity Equation

Figure 7.4 depicts a fixed volume in a tube where mass flows in and out. This situation follows the mass-conservation law, which states that what flows in and does not come out again will remain inside the volume since mass does not vanish.

The mass flowing in during a time interval, $\mathrm{d}t$, is

$$\mathrm{d}m_{\text{in}} = A\,\varrho_-\,\overbrace{v\,\mathrm{d}t}^{\mathrm{d}x}, \tag{7.20}$$

and the mass flowing out during the same time interval is

---

[3] The characteristic field impedance, Z_{w}, also known as wave impedance, will be introduced in Section 7.4

[4] We must check whether this linearization is still justified when the medium is flowing fast as it would in an exhaust system, and/or if we have rapid changes of v over x as would occur at area steps in a tube. It is worth noting that the second term is the more significant one in *flow dynamics* while the first term is typically neglected there

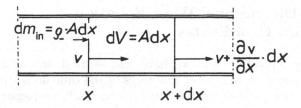

Fig. 7.4. A spatially-fixed differential volume, dV, of fluid in a tube

$$dm_{\text{out}} = A\,\varrho_-\,v\,dt + A\,\varrho_-\,\frac{\partial v}{\partial x}\,dx\,dt. \tag{7.21}$$

The difference, also known as the mass surplus, is

$$dm_\triangle = -A\,\varrho_-\,\frac{\partial v}{\partial x}\,dx\,dt. \tag{7.22}$$

If the mass inside the volume has changed, the mass density in the volume must have also changed, allowing us to write the mass surplus as

$$dm_\triangle = A\,dx\,\frac{\partial \varrho}{\partial t}\,dt. \tag{7.23}$$

Combining the two surplus expressions yields

$$-\varrho_-\,\frac{\partial v}{\partial x} = \frac{\partial \varrho}{\partial t}, \tag{7.24}$$

which is the *continuity equation*.

Since we want to use the field quantities p and v, we now substitute p by v by applying the state equation (7.10) as follows,

$$\partial \varrho_- = \frac{\partial p}{c^2} = \kappa_-\varrho_-\,\partial p. \tag{7.25}$$

By combining equations (7.24) and (7.25) we finally find the second linear differential wave equation, that is

$$-\frac{\partial v}{\partial x} = \kappa_-\,\frac{\partial p}{\partial t}. \tag{7.26}$$

Please note that κ has been replaced by κ_-, which is our *fourth linearization*[5].

---

[5] Since $\kappa_- \approx \kappa$ and $\varrho_- \approx \varrho$, we shall from now on omit the subscripts of κ_- and ϱ_- to simplify the notation

7.2 Three-Dimensional Wave Equation in *Cartesian* Coordinates

The two wave equations that we have just derived are valid for a one-dimensional axial wave in a tube. The walls of the tube do not play any role since we have a purely longitudinal wave that actually propagates in parallel to the walls.

Our equations are also valid for a bundle of tubes – shown in Fig. 7.5 – and nothing would change with respect to the wave if we took the tubes away. Consequently, the equations still hold for one-dimensional longitudinal waves in infinitely extended media.

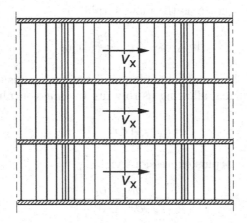

Fig. 7.5. Bundle of tubes with one-dimensional waves in them

Since there is no shear stress in fluids, waves from different directions in the medium just superimpose without influencing each other. This property allows us to formulate the three-dimensional wave equations in *Cartesian* coordinates by superimposing the wave components of the x, y, and z directions. By using the following three operators from vector analysis,

$$\overrightarrow{\mathrm{grad}\,p} = \overrightarrow{\nabla} p = \left\{ \frac{\partial p_x}{\partial x}; \frac{\partial p_y}{\partial y}; \frac{\partial p_z}{\partial z} \right\}, \quad \text{a vector,} \tag{7.27}$$

$$\mathrm{div}\,\overrightarrow{v} = \nabla\,\overrightarrow{v} = \left\{ \frac{\partial v_x}{\partial x} + \frac{\partial v_y}{\partial y} + \frac{\partial v_z}{\partial z} \right\}, \quad \text{a scalar, and} \tag{7.28}$$

$$\mathrm{div}\,\overrightarrow{\mathrm{grad}\,p} = \nabla^2 p = \left\{ \frac{\partial^2 p_x}{\partial x^2} + \frac{\partial^2 p_y}{\partial y^2} + \frac{\partial^2 p_z}{\partial z^2} \right\}, \quad \text{a scalar,} \tag{7.29}$$

we can write the following handsome set of linear acoustical wave equations for lossless fluids, with $\dot{v}$ and $\dot{p}$ denoting first-order time derivatives,

$$-\operatorname{grad} p = \varrho \, \dot{v}, \tag{7.30}$$

$$-\operatorname{div} v = \kappa \, \dot{p}, \tag{7.31}$$

where we have omitted the vector arrows for simplicity. Combination of (7.30) and (7.31) renders the wave equation with the second-order time derivative,

$$\nabla^2 p = \frac{1}{c^2}\ddot{p}. \tag{7.32}$$

The corresponding equations for the particle velocity, v, shown below, looks formally identical[6], namely,

$$\nabla^2 v = \frac{1}{c^2}\ddot{v}. \tag{7.33}$$

7.3 Solutions of the Wave Equation

Now we will present solutions of the wave equation. To keep the discussion simple, we restrict ourselves to one dimensional cases, specifically the wave equation in the following form, as predicted at the beginning of this chapter[7].

$$\frac{\partial^2 p}{\partial x^2} = \frac{1}{c^2}\frac{\partial^2 p}{\partial t^2}. \tag{7.34}$$

Two or three-dimensional solutions can be formed by superposition of one-dimensional solutions.

General Solution

This solution is known as the *d'Alembert* solution. *d'Alembert* assumes two waves propagating in opposite directions, x and $-x$. Both the forward progressing and returning waves have the same speed, c. The solution is written as

$$p(x,t) = p_{\rightarrow}\left(t - \frac{x}{c}\right) + p_{\leftarrow}\left(t + \frac{x}{c}\right). \tag{7.35}$$

Please note that, since this equation is a differential equation of the second order, it can be satisfied by any function $f(t \pm x/c)$ that is differentiable twice with respect to both time and space.

---

[6] In theoretical acoustics, the velocity potential, Φ, is frequently used. It is defined as $\vec{v} = -\overrightarrow{\operatorname{grad} \Phi}$. Its wave equation is $\nabla^2 \Phi = (1/c^2)\,\ddot{\Phi}$

[7] This form can also be derived by combining the differential of (7.19) by x with that of (7.26) by t

Solution for Harmonic Functions

This solution, called the *Bernoulli* solution, solution is a special case of *d'Alembert*'s solution applied to harmonic functions (sinusoids). Dealing with harmonic functions allows us to take advantage of complex notation by rewriting the wave equation as follows,

$$\frac{\partial^2 \underline{p}}{\partial x^2} - (j\beta)^2 \underline{p} = 0. \tag{7.36}$$

This is known as the *Helmholtz* form, in which

$$\beta = \frac{\omega}{c} = \omega \sqrt{\varrho \, \kappa}, \tag{7.37}$$

is called *phase coefficient*[8].

The solution can be found with the trial $\underline{p} = e^{\gamma x}$, which leads to the characteristic equation $\gamma^2 = (j\beta)^2$ and its solutions[9], $\gamma_{1,2} = \pm(j\beta)$. These solutions allow again for two waves in opposite directions, namely,

$$\underline{p}(x) = \underline{p}_{\rightarrow} e^{-j\beta x} + \underline{p}_{\leftarrow} e^{+j\beta x}. \tag{7.38}$$

Both the forward progressing and the returning waves go through a complete period for $\beta x_0 = 2\pi$. With x_0 equal to the *wavelength*, λ, we obtain the well-known formulae

$$\lambda f = c \quad \text{and} \quad \beta = \frac{\omega}{c} = \frac{2\pi}{\lambda}. \tag{7.39}$$

7.4 Field Impedance and Power Transport in Plane Waves

First we look at the forward progressing wave, expressed as

$$\underline{p}(x) = \underline{p}_{\rightarrow} e^{-j\beta x}, \quad \text{and} \quad \underline{v}(x) = \underline{v}_{\rightarrow} e^{-j\beta x}. \tag{7.40}$$

Following *Euler*, we obtain

$$\varrho \frac{\partial v}{\partial t} = -\frac{\partial p}{\partial x} \quad \text{and} \quad j\omega \underline{v}_{\rightarrow} e^{-j\beta x} \varrho = -(-j\beta) \underline{p}_{\rightarrow} e^{-j\beta x}, \tag{7.41}$$

and, finally, considering the returning way likewise,

---

[8] In physics this quantity is often called *wave number* or, more precisely, *angular wave number* and denoted by k. In fact, it is not a "number" since is has the dimension [1/length]

[9] γ is called *complex propagation coefficient*, also denoted complex wave number $\underline{k}$. We shall make use of it later, particularly, in Sections 8.3 and 11.2

$$Z_\text{w} = \frac{\underline{p}_{\rightarrow}\, e^{-j\beta x}}{\underline{v}_{\rightarrow}\, e^{-j\beta x}} = -\frac{\underline{p}_{\leftarrow}\, e^{+j\beta x}}{\underline{v}_{\leftarrow}\, e^{+j\beta x}} = \varrho\, c = \sqrt{\frac{\varrho}{\kappa}}. \qquad (7.42)$$

The real quantity, Z_w, is known as *characteristic field impedance* or *wave impedance*[10]. Since Z_w is real in our case, sound pressure and particle velocity are in phase, which generally holds for plane waves in lossless fluids.

The thus purely active (resistive) intensity, $\overline{I} = \text{Re}\{\underline{I}\}$, transported by the forward progressing wave is

$$\overline{I} = \frac{1}{2}\,\text{Re}\left\{\underline{p}_{\rightarrow}\, e^{-j\beta x}\left(\underline{v}_{\rightarrow}\, e^{-j\beta x}\right)^{*}\right\} = \frac{1}{2}\,|\underline{p}_{\rightarrow}|\,|\underline{v}_{\rightarrow}| = \frac{1}{2}\,Z_\text{w}|\underline{v}_{\rightarrow}|^{2}, \quad (7.43)$$

which also holds for the returning wave.

The transported active power, $\overline{P}$, results from multiplying the intensity with the area that it crosses perpendicularly, resulting in the following scalar product of two vectors, $\overline{P} = \overrightarrow{I} \cdot \overrightarrow{A_{\perp}}$.

7.5 Transmission-Line Equations and Reflectance

As previously mentioned, there is only axial wave propagation in tubes with much smaller diameters than wavelengths, $d \ll \lambda$.

Since the ratio of $\underline{p}$ and $\underline{v}$ inside the tube is only dependent on the terminating impedance of the tube, $\underline{Z}_0 = \underline{p}_0/\underline{v}_0$, it is convenient to formulate the solution of the wave equation in terms of the distance to the position of $\underline{Z}_0$. To this end, we substitute the coordinate x with $-l$ – shown in Fig. 7.6.

Now, because of $e^{\pm j\beta l}|_{l=0} = 1$ at position $l = 0$, we have $\underline{p}_0 = \underline{p}_{\rightarrow} + \underline{p}_{\leftarrow}$ and $\underline{v}_0 = \underline{v}_{\rightarrow} + \underline{v}_{\leftarrow}$. Keeping (7.42) in mind, we further have $Z_\text{w} = \underline{p}_{\rightarrow}/\underline{v}_{\rightarrow} = -\underline{p}_{\leftarrow}/\underline{v}_{\leftarrow}$, which leads to the expressions

$$Z_\text{w}\,\underline{v}_0 = \underline{p}_{\rightarrow} - \underline{p}_{\leftarrow} \quad \text{and} \quad \underline{p}_0/Z_\text{w} = \underline{v}_{\rightarrow} - \underline{v}_{\leftarrow}. \qquad (7.44)$$

From this point it follows that

$$\underline{p}_{\rightarrow} = \frac{1}{2}\left(\underline{p}_0 + Z_\text{w}\,\underline{v}_0\right), \qquad \underline{p}_{\leftarrow} = \frac{1}{2}\left(\underline{p}_0 - Z_\text{w}\,\underline{v}_0\right) \quad \text{and} \qquad (7.45)$$

$$\underline{v}_{\rightarrow} = \frac{1}{2}\left(\frac{\underline{p}_0}{Z_\text{w}} + \underline{v}_0\right), \qquad \underline{v}_{\leftarrow} = -\frac{1}{2}\left(\frac{\underline{p}_0}{Z_\text{w}} - \underline{v}_0\right). \qquad (7.46)$$

For an arbitrary position, $l = -x$, the following expressions result,

$$\underline{p}(l) = \frac{1}{2}\left(\underline{p}_0 + Z_\text{w}\,\underline{v}_0\right) e^{j\beta l} + \frac{1}{2}\left(\underline{p}_0 - Z_\text{w}\,\underline{v}_0\right) e^{-j\beta l} \quad \text{and} \qquad (7.47)$$

---

[10] If losses in the medium are to be considered, $\underline{Z}_\text{w}$ becomes necessarily complex with the addition of an imaginary component

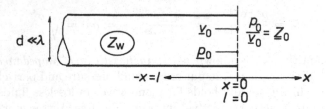

Fig. 7.6. One dimensional wave in a tube

$$\underline{v}(l) = \frac{1}{2}\left(\frac{\underline{p}_0}{Z_{\mathrm{w}}} + \underline{v}_0\right) \mathrm{e}^{\mathrm{j}\beta l} - \frac{1}{2}\left(\frac{\underline{p}_0}{Z_{\mathrm{w}}} - \underline{v}_0\right)\mathrm{e}^{-\mathrm{j}\beta l}. \tag{7.48}$$

Complex decomposition with $\mathrm{e}^{-\mathrm{j}\alpha} = \cos\alpha - \mathrm{j}\sin\alpha$ leads to the following two equations that are known as *transmission-line equations*[11],

$$\underline{p}(l) = \underline{p}_0\,\cos\beta l + \mathrm{j}Z_{\mathrm{w}}\underline{v}_0\,\sin\beta l, \tag{7.49}$$

$$\underline{v}(l) = \mathrm{j}\frac{\underline{p}_0}{Z_{\mathrm{w}}}\,\sin\beta l + \underline{v}_0\,\cos\beta l. \tag{7.50}$$

Another popular way of writing these equations is

$$\frac{\underline{p}(l)}{\underline{p}_0} = \cos\beta l + \mathrm{j}\frac{Z_{\mathrm{w}}}{\underline{Z}_0}\,\sin\beta l, \tag{7.51}$$

$$\frac{\underline{v}(l)}{\underline{v}_0} = \cos\beta l + \mathrm{j}\frac{\underline{Z}_0}{Z_{\mathrm{w}}}\,\sin\beta l. \tag{7.52}$$

Let us now discuss different terminations of the tube. Three cases are especially interesting.

- $\underline{Z}_0 = \underline{p}_0/\underline{v}_0 = Z_{\mathrm{w}}$ ... This means that there is no reflection at the terminal, and all power moves on. This case is called *wave match* and results in no returning wave. Consequently, (7.49) becomes

$$\underline{p}(l) = \underline{p}_0\,\mathrm{e}^{\mathrm{j}\beta l} \tag{7.53}$$

- $\underline{Z}_0 = \infty$ and, hence, $\underline{v}_0 = 0$ ... This case is called *hard termination* and is achieved by closing the tube with a rigid surface. In this case the forward progressing wave will be fully reflected and no power is able to leave through the terminal. Equations (7.49) and (7.50) render

$$\underline{p}(l) = \underline{p}_0\cos\beta l \quad \text{and} \quad \underline{v}(l) = \mathrm{j}\frac{\underline{p}_0}{Z_{\mathrm{w}}}\sin\beta l \tag{7.54}$$

This describes a so-called *standing wave*. The sound pressure varies at all positions sinusoidally as a function of time. Sound pressure and particle velocity are 90° out of phase

---

[11] This name originates from electrical engineering where the same equations hold but with p being replaced by u and v by i

- $\underline{Z}_0 = 0$ and, thus, $\underline{p}_0 = 0$ ... Now the end of the tube is open, resulting in *soft termination*. Neglecting any radiation out of the open end – which is of course idealizing – the sound pressure is zero on the terminating plane. The forward progressing and the returning wave will again superimpose to a standing wave but will have a different phase angle than the one in the hard-termination case. Again, there is no power leaving through the terminal

To describe the reflection at the terminal plane of the tube, the reflectance, $\underline{r}$, is a useful quantity. It is defined as

$$\underline{r} = \frac{\underline{p}_\leftarrow}{\underline{p}_\rightarrow} = \frac{\frac{1}{2}\left[\underline{p}_0 - Z_{\mathrm{w}}\underline{v}_0\right]}{\frac{1}{2}\left[\underline{p}_0 + Z_{\mathrm{w}}\underline{v}_0\right]} = \frac{\underline{Z}_0 - Z_{\mathrm{w}}}{\underline{Z}_0 + Z_{\mathrm{w}}}. \qquad (7.55)$$

We have $\underline{r} = 0$ for wave match, $\underline{r} = +1$ for hard termination and $\underline{r} = -1$ for soft termination. If intensity or power are under consideration, the following definitions[12] are used.

- $|r|^2$ ... degree of reflection
- $1 - |r|^2 = \alpha$ ... degree of absorption

7.6 The Acoustic Measuring Tube

The acoustic measuring tube, also known as *Kundt*'s tube, can be used to measure field impedances and reflectances in the terminal plane of a tube. Such a tube is schematically shown in Fig. 7.7. The tube can be excited by a sound source from one end, a sinusoidal sound in our case. The other end is terminated by the impedance to be measured. There is an arrangement of absorbing material in front of the source to avoid back reflection.

Standard Method

A microphone can be moved along the axis of the tube to measure the sound pressure as a function of its position. The sound pressure inside the tube follows the following equation, whereby $\underline{r} = |r|\,\mathrm{e}^{-\mathrm{j}\phi_{\mathrm{r}}}$,

$$\underline{p}(l) = \underline{p}_\rightarrow \left(\mathrm{e}^{+\mathrm{j}\beta l} + \underline{r}\,\mathrm{e}^{-\mathrm{j}\beta l}\right) = \underline{p}_\rightarrow \left(\mathrm{e}^{+\mathrm{j}\beta l} + |r|\mathrm{e}^{-\mathrm{j}(\beta l - \phi_{\mathrm{r}})}\right). \qquad (7.56)$$

For any $|r| \neq 0$, this function will have maxima and minima along l. In other words, a standing-wave behavior will show up – illustrated in the lower panel of Fig. 7.6. The maxima will be located at all positions of l where the phase angles of the two terms in (7.56) are equal, the minima accordingly, where the phase angles are opposite. The actual positions thus are,

---

[12] In the literature, these terms are often called reflection coefficient and absorption coefficient. However, their dimension is [1] and their value range is 0–1, or 0 %–100 %, respectively. This is why we prefer the term *degree*

- Maxima at $\beta l = -(\beta l - \phi_r) \pm n\pi$
- Minima at $\beta l = +(\beta l - \phi_r) \pm n\pi$

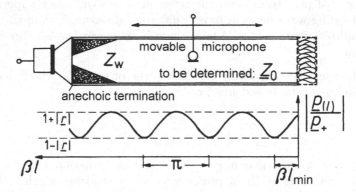

Fig. 7.7. *Kundt*'s tube for impedance measurement

This results in the following *standing-wave ratio*, S,

$$S = \left| \frac{p_{max}}{p_{min}} \right| = \frac{1 + |r|}{1 - |r|}, \tag{7.57}$$

from which we get the magnitude of the reflectance[13] as

$$|r|(l = 0) = \frac{S - 1}{S + 1}. \tag{7.58}$$

The phase angle of the reflectance, ϕ_r, can, for instance, be computed from the position of the first minimum, l_{min}, and results as

$$\phi_r(l = 0) = \beta \, l_{min}. \tag{7.59}$$

Transfer-Function Method

In an alternative approach, the so-called *transfer-function method*, the complex reflectance, $\underline{r}\,(l = 0)$, and, consequently, the terminating impedance, $\underline{Z}_0 = \underline{p}_0/\underline{v}_0$, can be measured without microphones having to be shifted mechanically. Instead, two microphones at a mutual distance of $l_\Delta = l_2 - l_1$ are employed at the positions 2 and 1 – depicted in Fig. 7.8. Starting from (7.49), we write[14]

---

[13] The theory of *Kundt*'s tube is homomorphic to the theory of the measuring line for electromagnetic waves. To convert reflectances into impedances, and vice versa, a graphical tool called *Smith* chart is useful. It converts an image of the complex $\underline{Z}$-plane onto an image of the complex $\underline{r}$-plane

[14] Note that all distances involved, that is, l_1, l_2 and $l_\Delta = l_2 - l_1$, have positive values

$$\underline{p}_2 = \underline{p}_1 \cos(\beta l_\Delta) + jZ_w \underline{v}_1 \sin(\beta l_\Delta) \quad \text{and, thus,} \tag{7.60}$$

$$\frac{\underline{p}_2}{\underline{p}_1} = \cos(\beta l_\Delta) + j\frac{Z_w}{\underline{Z}_1} \sin(\beta l_\Delta), \quad \text{or} \tag{7.61}$$

$$\underline{Z}_1 = \frac{jZ_w \sin(\beta l_\Delta)}{(\underline{p}_2/\underline{p}_1) - \cos(\beta l_\Delta)} . \tag{7.62}$$

With a straight-forward transformation along the distance $-l_2$, which follows from the transmission-line equations (7.51) and (7.52), $\underline{Z}_2$ can be transformed into the impedance at the surface of the probe to be tested, Z_0. The relevant formula is

$$\underline{Z}_0 = \frac{\underline{Z}_1 \cos(\beta l_1) + jZ_w \sin(\beta l_1)}{\cos(\beta l_1) + j(\underline{Z}_1/Z_w) \sin(\beta l_1)}, \tag{7.63}$$

In a similar way, we obtain a widely used formula for the complex reflectance, namely

$$\underline{r} = \frac{(\underline{p}_1/\underline{p}_2) - e^{-j\beta l_\Delta}}{e^{+j\beta l_\Delta} - (\underline{p}_1/\underline{p}_2)} e^{+j2\beta l_2} . \tag{7.64}$$

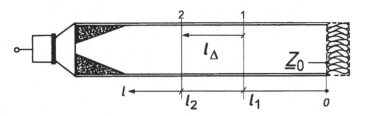

Fig. 7.8. Impedance measurement with the transfer-function method

The ratio $\underline{p}_1/\underline{p}_2 = \underline{T}_{pp}$ for a given frequency, ω_1, is a transfer factor. Its frequency function, $\underline{H}_{pp}(\omega)$, is called a *transfer function*. This fact lends the method its name.

The transfer-function method fails for frequencies where the nominator or denominator in (7.63) or (7.64) become zero. This is the case whenever $l_\Delta = \lambda/2$. To cover a wide frequency range, the condition $l_\Delta < \pi c/\omega$ must be fulfilled. To this end, more than two microphones at different positions may be employed – or one microphone that measures sequentially at different positions.

8

Horns and Stepped Ducts

The wave equations derived in the preceding chapter allow for calculation of arbitrary sound fields with any possible, physically meaningful boundary conditions. We had restricted ourselves to one-dimensional waves so far. These can, for instance, be observed in tubes with a diameter being small compared to the wavelength, that is $d \ll \lambda$. This condition guarantees that no other waveforms than axial ones can propagate in the tube. One-dimensional propagation also means that all wave planes perpendicular to the axial direction are planes of constant phase.

In the following, ducts shall be considered where the diameter varies with x. In other words, the area function $A = f(x)$ is no longer constant. Nevertheless, the condition $d \ll \lambda$ shall still holds. Two cases – depicted in Fig. 8.1 – will be discussed.

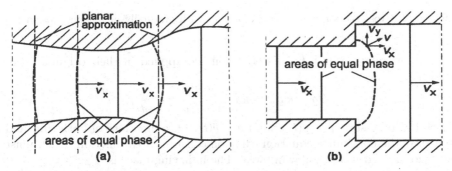

Fig. 8.1. Two types of ducts with non-constant area function. (a) continuous variation of the cross-sectional area, (b) stepped variations

- *Continuous variation* of the cross area. As long as this variation is only gradual compared to the wavelength, it is still justified to assume one-dimensional, axial propagation. Radial propagation can then be neglected

- *Step-like mutations* of the cross area as a function of x – so-called *stepped ducts*. Very close to the position where the step occurs, we certainly have radial components of the particle velocity. Yet, as radial waves cannot propagate, they can be neglected already at small distances away from the step. In fact, there we have plane waves again. If we look at the cross area just Δx in front of the step and again Δx behind it, we can state that the axial component of the volume velocity, $\underline{q} = A \, \underline{v}$, is the same in both cross sections. Thus, in our calculations, we can neglect any *modal dispersion* in the immediate vicinity of the step position and set the volume velocity at both sides of the step to equal

8.1 *Webster*'s Differential Equation – the Horn Equation

This chapter deals with the condition where the area function varies only gradually, and perpendicular areas are areas of approximately constant phase. This case is captured by the so-called *Webster equation* or *Horn equation*. Figure 8.2 illustrates the derivation of this differential equation. Please consider the elementary volume between the cross areas at x and $x+dx$.

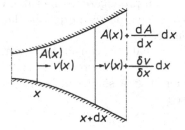

Fig. 8.2. Cross-section of a *horn*, i.e. a duct with gradually increasing cross area

The state equation and *Euler*'s equation are applied in their original form, namely,

$$\partial\varrho = \kappa\,\varrho\,\partial p \quad \text{and} \quad -\frac{\partial p}{\partial x} = \varrho\,\frac{\partial v}{\partial t}\,, \tag{8.1}$$

whereby, to be sure, $p = p(t,x)$ and $v = v(t,x)$ are functions of both time and space. In the continuity equation, the non-constant area function, A(x), must be reconsidered in the following way. The inflowing mass is

$$dm_{\text{in}} = \varrho A(x)\, v\, dt\,. \tag{8.2}$$

The outflowing mass is

$$dm_{\text{out}} = \varrho \left[A(x) + \frac{\mathrm{d}A}{\mathrm{d}x}\,\mathrm{d}x \right] \left(v + \frac{\partial v}{\partial x}\,\mathrm{d}x \right) \mathrm{d}t =$$

$$\varrho \left[A(x)\,v + A(x)\,\frac{\partial v}{\partial x}\,\mathrm{d}x + v\,\frac{\mathrm{d}A}{\mathrm{d}x}\,\mathrm{d}x + \overbrace{(\cdots\cdots)}^{\text{2}^{\text{nd}}\text{ order differentials}} \right] \mathrm{d}t . \qquad (8.3)$$

Neglecting the second-order differentials in the sum, we get the mass surplus as follows,

$$dm_\Delta = -\varrho\,A(x) \left(\frac{\partial v}{\partial x} + \frac{1}{A(x)}\frac{\partial A}{\partial x}\,v \right) \mathrm{d}t\,\mathrm{d}x = \overbrace{\left(A(x) + \frac{\mathrm{d}A(x)}{2} \right)}^{\text{average area}} \mathrm{d}x\,\frac{\partial \varrho}{\partial t}\,\mathrm{d}t . \qquad (8.4)$$

Again neglecting second-order differentials consequently yields

$$-\varrho \left(\frac{\partial v}{\partial x} + \frac{1}{A(x)}\frac{\mathrm{d}A}{\mathrm{d}x}\,v \right) = \frac{\partial \varrho}{\partial t} , \qquad (8.5)$$

which is the *modified continuity equation*.

To get p as the second field quantity instead of ϱ, the state equation (8.1) is used and we arrive at

$$-\left(\frac{\partial v}{\partial x} + \frac{1}{A(x)}\frac{\mathrm{d}A}{\mathrm{d}x}\,v \right) = \kappa\,\frac{\partial p}{\partial t} . \qquad (8.6)$$

Combining (8.1) and (8.6) leads to *Webster's* equation, which is

$$\frac{\partial^2 p}{\partial x^2} + \left[\frac{1}{A(x)}\frac{\mathrm{d}A}{\mathrm{d}x} \right]\frac{\partial p}{\partial x} = \frac{1}{c^2}\frac{\partial^2 p}{\partial t^2} . \qquad (8.7)$$

Please note that the term $[1/A(x)]\,(\mathrm{d}A/\mathrm{d}x)$ is identical to $\mathrm{d}[\ln A(x)]/\mathrm{d}x$. For $A(x) = \text{const}$, *Webster's* equation reduces to the normal one-dimensional wave equation[1].

For a number of analytically defined area functions, *Webster's* equation can be integrated in closed form. In the following section, we take two of them as examples, namely, conical and exponential horns.

8.2 Conical Horns

For the conical horn – sketched in Fig. 8.3 – the area function is

$$A(x) = A_0 \left(\frac{x}{x_0} \right)^2 . \qquad (8.8)$$

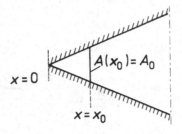

Fig. 8.3. Conical horn

Inserting the area function into *Webster*'s equation results in

$$\frac{\partial^2 p}{\partial x^2} + \frac{2}{x}\frac{\partial p}{\partial x} = \frac{1}{c^2}\frac{\partial^2 p}{\partial t^2}, \tag{8.9}$$

whereby we have used

$$\frac{1}{A(x)}\frac{\mathrm{d}A}{\mathrm{d}x} = \frac{A_0}{A(x)}2\left(\frac{x}{x_0}\right)\frac{1}{x_0} = \frac{2}{x}. \tag{8.10}$$

Equation (8.9) can also be written in the following form as can be proven by differentiating,

$$\frac{\partial^2(p\,x)}{\partial x^2} = \frac{1}{c^2}\frac{\partial^2(p\,x)}{\partial t^2}. \tag{8.11}$$

By inspecting this formula, it becomes obvious that its form corresponds to the one-dimensional wave equation, yet, instead of p, we now have a product $p\,x = g$. The solution is approached in the well known way by

$$\underline{g}(x) = \underline{p}(x)\,x = \underline{g}_{\rightarrow}\mathrm{e}^{-\mathrm{j}\beta x} + \underline{g}_{\leftarrow}\mathrm{e}^{+\mathrm{j}\beta x}. \tag{8.12}$$

By restricting ourselves to the forward progressing (outbound) wave, we get the following results for p and v,

$$\underline{p}_{\rightarrow}(x) = \frac{\underline{g}_{\rightarrow}}{x}\mathrm{e}^{-\mathrm{j}\beta x} \quad \text{and} \tag{8.13}$$

$$\underline{v}_{\rightarrow}(x) = \underline{g}_{\rightarrow}\left[\frac{1}{\varrho\,c\,x} + \frac{1}{\mathrm{j}\omega\,\varrho\,x^2}\right]\mathrm{e}^{-\mathrm{j}\beta x}, \tag{8.14}$$

where the solution for the particle velocity, v, has been found via the solution for p by applying *Euler*'s equation (8.1) with $\beta = \omega/c$ as follows,

$$-\underline{g}_{\rightarrow}(x)\left[-\frac{1}{x^2}\mathrm{e}^{-\mathrm{j}\beta x} - \mathrm{j}\beta\mathrm{e}^{-\mathrm{j}\beta x}\frac{1}{x}\right] = \mathrm{j}\omega\,\varrho\,\underline{v}. \tag{8.15}$$

The sound-field from the conical horn can be divided into a *near field* and a *far field*. The threshold between the two is defined as the position where

---

[1] *Webster*'s equation for v looks different from that as derived above for p

the magnitudes of the real and imaginary parts of the velocity are just equal, which is at

$$\left|\frac{1}{\varrho c x_{\mathrm{ff}}}\right| = \left|\frac{1}{j\omega \varrho x_{\mathrm{ff}}^2}\right| , \qquad (8.16)$$

resulting in a far-field distance of

$$x_{\mathrm{ff}} = \frac{\lambda}{2\pi} = \frac{1}{\beta_{\mathrm{ff}}} = \frac{c}{\omega_{\mathrm{ff}}} , \qquad (8.17)$$

with $x < \lambda/2\pi$ being the *near field* and $x > \lambda/2\pi$ the *far field*.

The sound pressure, p, decreases to half with a doubling of the distance, x. In other words, the decrease is 6 dB per distance doubling. For v the situation is more complicated. In the near field, v decreases with $1/x^2$ per distance doubling, which is 12 dB decrease per distance doubling. However, in the far field, v behaves like p with 6 dB decrease per distance doubling.

The field impedance $\underline{Z}_{\mathrm{f}}$ of the conical sound field results from dividing (8.13) by (8.14) and is

$$\underline{Z}_{\mathrm{f}}(x) = \frac{\underline{p}_{\rightarrow}(x)}{\underline{v}_{\rightarrow}(x)} = \frac{1}{\frac{1}{\varrho c} + \frac{1}{j\omega\varrho x}} = \varrho c \, \frac{j\frac{2\pi x}{\lambda}}{1 + j\frac{2\pi x}{\lambda}} . \qquad (8.18)$$

For this field impedance we can draw a substitute, namely, a long tube with a short branch where a concentrated mass is positioned – depicted in Fig. 8.4. The right panel of the figure shows an electro-acoustic analogy.

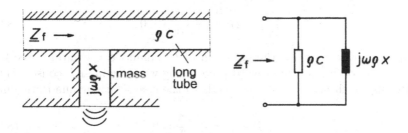

Fig. 8.4. Equivalent circuits for conical horns

The reactive (imaginary) component, $j\omega \varrho x$, is the so-called *co-vibrating medium mass*. This component swings about without transporting active power. The active (real) component ϱc becomes relatively (not absolutely!) stronger with increasing distance. For $x \gg \lambda/2\pi$, $\underline{Z}_{\mathrm{f}}$ approaches ϱc. Note that ϱc is the field impedance in a tube with a constant diameter and, thus, the specific field impedance of the medium, Z_{w}.

8.3 Exponential Horns

The area function of the exponential horn – see Fig. 8.5 – is given by

$$A(x) = A_0\, e^{2\epsilon x}, \tag{8.19}$$

with $\epsilon > 0$ being the so-called *flare coefficient*.

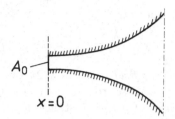

Fig. 8.5. Exponential horn

By differentiation we get

$$\frac{1}{A(x)}\frac{\mathrm{d}A}{\mathrm{d}x} = \frac{\mathrm{d}\left[\ln A(x)\right]}{\mathrm{d}x} = 2\epsilon \quad \text{and, thus,} \tag{8.20}$$

$$\frac{\partial^2 p}{\partial x^2} + 2\epsilon\frac{\partial p}{\partial x} = \frac{1}{c^2}\frac{\partial^2 p}{\partial t^2}\,. \tag{8.21}$$

The structure of this equation can most easily be understood by applying complex notation, which leads to

$$\frac{\delta^2 \underline{p}}{\partial x^2} + 2\epsilon\frac{\partial \underline{p}}{\partial x} + \frac{\omega^2}{c^2}\underline{p} = 0\,, \tag{8.22}$$

an equation which recalls the equation of the damped oscillator – see Section 2.3. We confine to the forward progressing wave again and, consequently, try the approach $\underline{p}(x) = e^{\underline{\gamma}x}$ that leads to the characteristic quadratic equation

$$\underline{\gamma}^2 + 2\epsilon\underline{\gamma} + \frac{\omega^2}{c^2} = 0 \tag{8.23}$$

with its two solutions

$$\underline{\gamma}_{1,2} = -\epsilon \pm \sqrt{\epsilon^2 - \frac{\omega^2}{c^2}} = -\epsilon \pm \mathrm{j}\sqrt{\frac{\omega^2}{c^2} - \epsilon^2}\,. \tag{8.24}$$

The complex quantity, $\underline{\gamma}$, is termed *propagation coefficient*, whereby

$$\underline{\gamma} = \alpha + \mathrm{j}\beta\,, \tag{8.25}$$

with α being the *damping coefficient* and β the *phase coefficient*.
The solution of the wave equation, hence, is an exponential function decreasing with x. This is called *spatial damping*. The general solutions for p and v in the forward progressing wave can be formulated as follows,

$$\underline{p}_{\rightarrow}(x) = \underline{p}_{\rightarrow}(0)\, e^{-\epsilon x} e^{-j\left(\sqrt{\frac{\omega^2}{c^2} - \epsilon^2}\right)x} = \underline{p}_{\rightarrow}(0)\, e^{-\alpha x} e^{-j\beta x} \quad \text{and} \quad (8.26)$$

$$\underline{v}_{\rightarrow}(x) = \frac{\epsilon + j\sqrt{\frac{\omega^2}{c^2} - \epsilon^2}}{j\omega\varrho}\, \underline{p}_{\rightarrow}(x). \tag{8.27}$$

Again, the solution for v has been derived from the one for p by applying *Euler*'s equation (8.1).

A prerequisite for wave propagation is that the expression under the square root is positive and, thus, results in a phase coefficient, β. This is the case when $\omega^2/c^2 > \epsilon^2$ and, accordingly, $2\pi/\lambda > \epsilon$ holds. In fact, this is fulfilled above a limiting frequency

$$\omega_l = \epsilon\, c. \tag{8.28}$$

Below ω_l, there is an exponential fade-out as the expression under the root becomes negative, and we then end up with pure damping without wave propagation. Physically, this means that mass is shifted about, but no energy is transported as no sufficient compression takes place. ω_l decreases with decreasing flare coefficient, ϵ. In other words, the slimmer the horn, the lower the limiting frequency.

Please note that the phase velocity, c_{ph}, in the exponential horn, is different from that in a free plane wave, c, namely,

$$c_{\mathrm{ph}} = \frac{\omega}{\beta} = \frac{\omega}{\sqrt{\frac{\omega^2}{c^2} - \epsilon^2}}. \tag{8.29}$$

Furthermore, c_{ph} is frequency-dependent. This effect is called *dispersion* since different frequency components travel with different speed and, thus, the different wave components arrive at the end of the horn at different instances[2].

The so-called *group-delay distortions*, which describe the frequency-dependent delay of the envelope of a transmitted signal, are highest close to the limiting frequency. The group delay, τ_{gr}, over a wave-traveling distance of l is in our case

$$\tau_{\mathrm{gr}} = \frac{d\beta}{d\omega} = \frac{l}{c\sqrt{1 - (\frac{\omega_l}{\omega})^2}}, \tag{8.30}$$

The field impedance in the exponential horn, $\underline{Z}_{\mathrm{f}}$, is given by

$$\underline{Z}_{\mathrm{f}} = \frac{\underline{p}_{\rightarrow}}{\underline{v}_{\rightarrow}} = \frac{j\omega\varrho}{\epsilon + j\sqrt{\frac{\omega^2}{c^2} - \epsilon^2}} = \varrho c \left[\sqrt{1 - (\frac{\omega_l}{\omega})^2} + j(\frac{\omega_l}{\omega}) \right]. \tag{8.31}$$

As with the conical horn, $\underline{Z}_{\mathrm{f}}$ approaches $\varrho c = Z_{\mathrm{w}}$ with increasing frequency since we have $\underline{Z}_{\mathrm{f}} \Rightarrow \varrho c$ for $\omega \gg \omega_l$.

---

[2] This, by the way, contributes to the characteristic timbre of horn loudspeakers.

8.4 Radiation Impedances and Sound Radiation

The acoustic power that is sent out by an electro-acoustic transducer or any other sound source, is proportional to the real part of the impedance, $r_{rad} = \mathrm{Re}\{\underline{Z}_{rad}\}$ that terminates the source at its acoustic output port. Since this impedance is formed by the sound field coupled to the source, we call it *radiation impedance*, $\underline{Z}_{rad}$, and its real part *radiation resistance*, r_{rad}. The radiation impedance is a mechanic impedance – refer to Section 4.5 – namely,

$$\underline{Z}_{rad} = \frac{\underline{F}}{\underline{v}} . \tag{8.32}$$

The radiated power, then, is

$$\overline{P} = \frac{1}{2} \, \mathrm{Re}\{\underline{Z}_{rad}\} \, |\underline{v}|^2 = \frac{1}{2} r_{rad} \, |\underline{v}|^2 . \tag{8.33}$$

The following relation holds between the field impedance, $\underline{Z}_f$, and the radiation impedance, $\underline{Z}_{rad}$,

$$\underline{Z}_{rad} = \int_A \underline{Z}_f \, dA , \tag{8.34}$$

with A being the effective radiation area.

For transducers that radiate into a horn, the effective area is equal to the area of the mouth of the horn in the optimal case. In the synopsis shown in Fig. 8.6, we assume that the tube/horn is so long that no waveforms are reflected back from the opening, but that the diameter is still small as compared to the wavelength, that is $d \ll \lambda$. This is, of course, an idealized assumption.

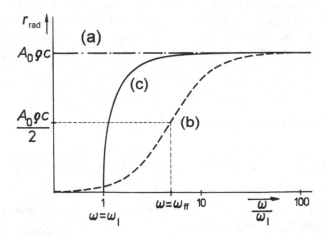

Fig. 8.6. Schematic plot of the radiation resistance of **(a)** a tube, **(b)** a conical horn, and **(c)** an exponential horn. Frequencies normalized to the limiting frequency, ω_l, of the exponential horn

For the tube with a constant cross section, we get

$$r_{rad} = Z_w A_0 = \varrho c A_0 \,, \tag{8.35}$$

for the conical horn

$$r_{rad}(A_0) = A_0 \operatorname{Re}\{\underline{Z}_f\} = A_0 \, \varrho c \, \frac{(\frac{\omega}{c} x_0)^2}{1 + (\frac{\omega}{c} x_0)^2} \,, \tag{8.36}$$

and for the exponential horn

$$r_{rad}(A_0) = A_0 \operatorname{Re}\{\underline{Z}_f\} = A_0 \, \varrho c \, \sqrt{1 - (\frac{\omega_l}{\omega})^2} \,. \tag{8.37}$$

For the conical horn, ω_{ff}, which forms the threshold between near- and farfield at a given distance from the mouth, x_1, is independent of the opening angle of the horn. For the exponential horn, however, the limiting frequency, ω_l depends on the flare coefficient, ϵ.

The exponential horn is, among all horns that can be described with *Webster*'s equation, the one with the steepest increase of $\operatorname{Re}\{\underline{Z}_{rad}\} = r_{rad}$ as a function of frequency. However, by considering the curvature of the waves, one can find even more advantageous forms – for example, spherical-wave horns.

8.5 Steps in the Area Function

We shall now discuss the situation at the position of the step in a tube – shown in Fig. 8.7. Left and right of the step, we have tubes with constant, though different diameters. As already mentioned at the beginning of this chapter, perpendicular modes at this position may be neglected because they cannot propagate as long as $d \ll \lambda$ holds at both sides of the step.

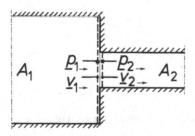

Fig. 8.7. Tube with steps in the area function

This means that, slightly away from the step, we only have axial waves again.

The boundary conditions, thus, are

$$\underline{p}_1 = \underline{p}_2, \quad \text{and} \tag{8.38}$$

$$A_1 \underline{v}_1 = A_2 \underline{v}_2, \quad \text{which is} \quad \underline{q}_1 = \underline{q}_2. \tag{8.39}$$

This means that both quantities are taken as continuous at the step. At the step a reflected wave is created. Accordingly, by combining

$$\underline{p}_{1\rightarrow} + \underline{p}_{1\leftarrow} = \underline{p}_{2\rightarrow} \quad \text{with} \quad A_1 \left(\frac{\underline{p}_{1\rightarrow}}{Z_\mathrm{w}} - \frac{\underline{p}_{1\leftarrow}}{Z_\mathrm{w}} \right) = \frac{\underline{p}_{2\rightarrow}}{Z_\mathrm{w}} A_2, \tag{8.40}$$

we get a reflectance

$$\underline{r} = \frac{\underline{p}_{1\leftarrow}}{\underline{p}_{1\rightarrow}} = \frac{A_1 - A_2}{A_1 + A_2}. \tag{8.41}$$

As $\underline{q}$ is continuous at the step, it makes sense to introduce this quantity to deal with stepped-duct problems rather than the particle velocity $\underline{v}$. The transmission-line equations (7.49) and (7.50) can, thus, be rewritten with $\underline{q}$ instead of $\underline{v}$ as follows,

$$\underline{p}(l) = \underline{p}_0 \cos \beta l + j Z_\mathrm{L} \underline{q}_0 \sin \beta l, \quad \text{and} \tag{8.42}$$

$$\underline{q}(l) = j \frac{\underline{p}_0}{Z_\mathrm{L}} \sin \beta l + \underline{q}_0 \cos \beta l, \tag{8.43}$$

where

$$Z_\mathrm{L} = \frac{Z_\mathrm{w}}{A} = \frac{1}{A} \sqrt{\frac{\varrho}{\kappa}} = \sqrt{\frac{m'_\mathrm{a}}{n'_\mathrm{a}}}, \tag{8.44}$$

is the specific acoustic impedance of the respective tube. m'_a is the acoustic mass per length, the *mass load*, n'_a is the acoustic compliance per length, the *compliance load*.

The two relevant energies now also come out as length-related quantities, namely, kinetic-energy per length,

$$W' = \frac{1}{2} m'_\mathrm{a} q^2, \tag{8.45}$$

and potential-energy per length,

$$W' = \frac{1}{2} n'_\mathrm{a} p^2. \tag{8.46}$$

The reflectance at the step between two tubes, each with constant cross-section, results as

$$\underline{r} = \frac{Z_{\mathrm{L}_2} - Z_{\mathrm{L}_1}}{Z_{\mathrm{L}_2} + Z_{\mathrm{L}_1}}. \tag{8.47}$$

Please note that by taking $\underline{p}$ analogous to $\underline{u}$, and $\underline{q}$ for $\underline{i}$, we see a complete analogy to the electric transmission line where we observe $\underline{i}_1 = \underline{i}_2$ and $\underline{u}_1 = \underline{u}_2$ at steps.

8.6 Stepped Ducts

With $\underline{q}$ as the second wave quantity, we can easily deal with stepped ducts by means of electric analogies. Furthermore, we can include acoustic concentrated elements into our consideration within the same analogue circuits – refer to Section 2.6. This means, in fact, that the theories of analysis and synthesis of electric networks, including transmission lines with and without losses, can be directly used to deal with acoustical problems. For example, acoustic filters with pre-described transfer functions can be designed in this way, including high-pass, low-pass and band-pass filters. This possibility is, for example, exploited in the design of mufflers for car-exhaust systems.

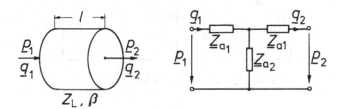

Fig. 8.8. Equivalent circuit for an acoustic tube segment

For the application of this method it is worth recalling that sections of transmission lines and, consequently, acoustic tubes, can be described by T-equivalents as given in the following matrix equation – see Fig. 8.8.

$$\begin{pmatrix} \underline{p}_1 \\ \underline{q}_1 \end{pmatrix} = \begin{pmatrix} \cos \beta l & jZ_L \sin \beta l \\ j\frac{1}{Z_L} \sin \beta l & \cos \beta l \end{pmatrix} \begin{pmatrix} \underline{p}_2 \\ \underline{q}_2 \end{pmatrix}. \tag{8.48}$$

These are the so-called two-port equations of a tube section, formulated in wave-parameter form. Two-port theory says that the following relations holds,

$$1 + \frac{\underline{Z}_{a_1}}{\underline{Z}_{a_2}} = \cos \beta l \quad \text{and} \quad \frac{1}{\underline{Z}_{a_2}} = j\frac{1}{Z_L} \sin \beta l, \tag{8.49}$$

further,

$$\underline{Z}_{a_1} = jZ_L \tan \frac{\beta l}{2} \quad \text{and} \quad \underline{Z}_{a_2} = -jZ_L \frac{1}{\sin \beta l}. \tag{8.50}$$

Please note that because transcendental functions (tan, sin) are involved, the elements can in principle not be realized by concentrated acoustic elements. Yet, for sections of small lengths, that is $l \ll \lambda$, the following approximations apply,

$$\tan \frac{\beta l}{2} \approx \frac{\beta l}{2} \quad \text{and} \quad \frac{1}{\sin \beta l} \approx \frac{1}{\beta l}. \tag{8.51}$$

Thereupon, with

$$Z_\mathrm{L} = \frac{1}{A}\sqrt{\frac{\varrho}{\kappa}} \quad \text{and} \quad \beta = \omega\sqrt{\varrho\kappa}, \tag{8.52}$$

we get

$$\underline{Z}_{a_1} \approx \frac{1}{2}\,\mathrm{j}\omega\,\frac{\varrho}{A}\,l = \frac{1}{2}\,\mathrm{j}\omega\,m'_a\,l \quad \text{and} \quad \underline{Z}_{a_2} \approx \frac{1}{\mathrm{j}\omega\,\kappa\,A\,l} = \frac{1}{\mathrm{j}\omega\,n'_a\,l}. \tag{8.53}$$

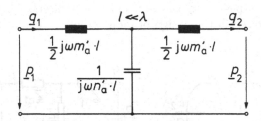

Fig. 8.9. Equivalent circuit for a tube segment

Figure 8.9 shows an equivalent circuit for a short section of a tube. This equivalent circuit is, for example, in use to calculate the transfer function of the human vocal tract or ear canal. The principle is depicted in Fig. 8.10. The higher the attempted accuracy of the calculation, the more sections have to be assumed.

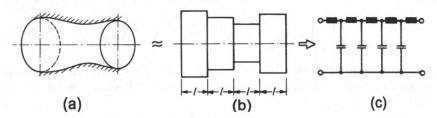

Fig. 8.10. Approximation of a tube with varied cross-sectional area

Finally, in this section, we shall treat the case of a very short narrowing or widening in a tube with otherwise constant cross section. The widening acts like a concentrated, branching spring, Δn, the narrowing like a concentrated serial mass, Δm. This can be figurately conceptualized as follows, taking a widening section with the length $l_2 \ll \lambda$ as example – Fig. 8.11(a).

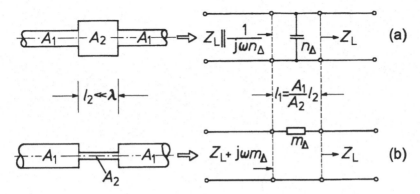

Fig. 8.11. Equivalent circuit for short tube segments, **(a)** short widening in a long constant-diameter tube, **(b)** short narrowing

From (8.53) we learn that

$$m'_{a_1} = \frac{A_2}{A_1} m'_{a_2} \quad \text{and} \quad n'_{a_1} = \frac{A_1}{A_2} n'_{a_2} . \tag{8.54}$$

Now, the widening section with the length contains the mass

$$m_a = m'_{a_2} l_2 . \tag{8.55}$$

If this mass were loaded upon the constant-diameter tube, we would need the length

$$l_1 = \frac{A_1}{A_2} l_2 . \tag{8.56}$$

Yet, a section of cross section A_1 and length l_1 would have a compliance of

$$n_a = \frac{A_2}{A_1} n'_{a_2} l_1 . \tag{8.57}$$

What we actually have at the widening section, is, however,

$$n_a = n_{a_2} l_2 = \left(\frac{A_1}{A_2} \right) n_{a_1} \left(\frac{A_1}{A_2} \right) l_1 = \left(\frac{A_1}{A_2} \right)^2 n_{a_1} l_1 . \tag{8.58}$$

In other words, the widening section acts like a section of the constant-diameter tube that has a cross section of A_1 and a length of

$$l_1 = \frac{A_1}{A_2} l_2 , \tag{8.59}$$

plus an additional parallel spring of

$$n_\Delta = \left[\left(\frac{A_2}{A_1} \right)^2 - 1 \right] n_a . \tag{8.60}$$

For the additional serial mass for a narrowing section, m_Δ – see Fig. 8.11(b) – the explanation would run accordingly.

9

Spherical Sound Sources and Line Arrays

The wave equation, $\nabla^2 p = \ddot{p}/c^2$ as derived in Sections 7.1 and 7.2 theoretically determines all possible sound fields in idealized fluids, that is, gases and liquids. The special task of computing sound fields for particular cases requires solutions of the wave equation for particular boundary conditions. In general, this task can be mathematically expensive, but there are helpful computer programs available, some of which are based on numerical methods like the finite-element method, FEM, or the boundary-element method, BEM. In praxi, approximations are often sufficient to understand the structure of a problem.

Closed solutions of the wave equation only exist for a limited number of special cases. We have already introduced the plane wave as one-dimensional solution in *Cartesian* coordinates. A few further one-, two- and three-dimensional cases are solvable in closed form, especially when symmetries allow for simplified formulations using appropriate coordinate systems as is the case for spherical or cylindrical coordinates.

In this chapter, we shall discuss basic solutions of the wave equation in spherical coordinates. In the same way that periodical time signals can be decomposed into *Fourier* harmonics, spherical sound waves can be decomposed into *spherical harmonics*[1].

In order to start with the essential basics, we focus on the spherical harmonics of 0th and 1st order and the sound sources that emit them. This also makes sense from the engineering standpoint since 0th and 1st order sound sources are of great practical relevance, mainly for the following two reasons.

---

[1] Spherical harmonics are *eigen-functions* of the wave equation in spherical coordinates

- At low frequencies many sound emitters act approximately like sources of 0th or 1st order spherical waves

- According to the principle of *Huygens*, each point on a wave front can be considered to be the origin of a spherical wave. Many sound fields can be estimated in a comparatively simple way by using this principle. Sound radiators with arbitrary directional characteristics can also be synthesized from spherical sound waves

9.1 Spherical Sound Sources of 0th Order

The wave equation allows for a one-dimensional, point-symmetric solution. This is a sound wave where all parameters only depend on the distance from the origin, r. The solution does not depend on the direction of propagation, which is always radial and directed either outward or toward the center. This type of wave is called a spherical wave of the 0th order, and a sound source that emits such a wave is called a spherical source of 0th order.

To derive the appropriate wave equation, it is helpful to transform the wave equation from *Cartesian* coordinates, x, y, z, into spherical coordinates, ϕ, δ, r. This is accomplished with the following well known formula,

$$
\begin{aligned}
\Delta = \nabla^2 &= \left[\frac{\partial^2}{\partial x^2} + \frac{\partial^2}{\partial y^2} + \frac{\partial^2}{\partial z^2} \right] \\
&= \left[\frac{1}{r^2} \frac{\partial}{\partial r} \left(r^2 \frac{\partial}{\partial r} \right) + \frac{1}{r^2 \sin \delta} \frac{\partial}{\partial \delta} \left(\sin \delta \frac{\partial}{\partial \delta} \right) + \frac{1}{r^2 \sin^2 \delta} \frac{\partial^2}{\partial \varphi^2} \right] . \quad (9.1)
\end{aligned}
$$

Because the assumed sound field is point symmetric and only changes in the radial direction, we can state that

$$
\frac{\partial}{\partial \delta} = \frac{\partial}{\partial \varphi} \equiv 0 . \quad (9.2)
$$

This leads to the wave equation for the 0th order spherical wave,

$$
\frac{1}{r^2} \frac{\partial}{\partial r} \left(r^2 \frac{\partial}{\partial r} \right) p = \frac{\partial^2 p}{\partial r^2} + \frac{2}{r} \frac{\partial p}{\partial r} = \frac{1}{c^2} \frac{\partial^2 p}{\partial t^2} . \quad (9.3)
$$

Note that this equation is identical to the wave equation for conical horns, which we derived in Section 8.2. The only difference is that x has been replaced by r. This congruence is intuitively plausible since the spherical wave can be thought of as a sound field composed of an infinite number of adjacent, very slim conical horns. This is illustrated in Fig. 9.1. If the walls of these conical horns are removed, the sound field stays the same since there is only radial propagation.

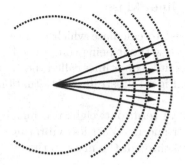

Fig. 9.1. Spherical waves of 0th order as a composition of conical waves

The solutions for the outward-progressing wave in the 0th-order spherical sound field are

$$\underline{p}_{\to}(r) = \frac{\underline{g}_{\to}}{r}\, e^{-j\beta r} \quad \text{and} \tag{9.4}$$

$$\underline{v}_{\to}(r) = \underline{g}_{\to} \left[\frac{1}{\varrho c\, r} + \frac{1}{j\omega\, \varrho r^2} \right] e^{-j\beta r}. \tag{9.5}$$

The field impedance of the diverging wave is

$$\underline{Z}_{\text{f}} = \varrho c\, \frac{j\frac{2\pi}{\lambda}}{1 + j\frac{2\pi r}{\lambda}} = \frac{1}{\frac{1}{\varrho c} + \frac{1}{j\omega \varrho r}}. \tag{9.6}$$

The near field is $r < \lambda/2\pi$, and the far field is $r > \lambda/2\pi$.

Spherical sound fields of the 0th order are radiated by spherical sound sources of 0th order, also called *breathing spheres* – shown in Fig. 9.2 (a).

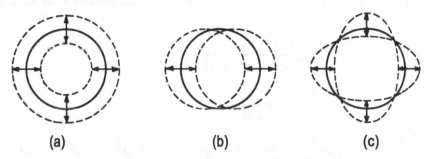

(a)   (b)   (c)

Fig. 9.2. Sound sources for spherical waves. (a) 0th-order source, also called *breathing* sphere, (b) an example of 1st-order sources, also called rigid *oscillating spheres*, (c) an example of 2nd-order sources. NB: There are $2n+1$ possible modes per order, n, with $n = 0, 1, 2 \ldots$

The Co-vibrating Medium Mass

It is an interesting exercise to calculate which part of the near field medium mass moves back and forth without being compressed. Because this part does not transmit active power, it is sometimes called the *Watt*-less-vibrating mass. The equivalent circuit in Fig 8.4 illustrates this situation. The diameter of the breathing sphere is r_0.

In the far field the real term outweighs the imaginary one. As a result, there is no reactive power and no *Watt*-less vibrating mass. In the near field the particle velocity flows through the mass so that

$$\left|\frac{\underline{p}}{\underline{v}}\right| = \omega \varrho r_0, \tag{9.7}$$

' and therefore, by implementing *Newton*'s law,

$$\left|\frac{\underline{F}}{\underline{v}}\right| = \omega m = \omega \varrho r_0 A_0. \tag{9.8}$$

By inserting the formula for the area of the sphere, $A_0 = 4\pi r_0^2$, the co-vibrating mass is found to be

$$m_{co} = 4\pi r_0^3 \varrho. \tag{9.9}$$

This is three times the mass of the medium inside the sphere if the medium is the same inside and outside.

Radiated Active Power and Source Strength

The radiated active power of a 0th-order spherical sound source is as follows – refer to Section 8.4,

$$\overline{P} = \frac{1}{2} \overbrace{A(r)\,\text{Re}\{\underline{Z}_{\text{f}}(r)\}}^{\text{radiation resistance},\, r_{\text{rad}}} |\underline{v}(r)|^2$$

$$= \frac{1}{2} \varrho c \frac{\left(\frac{\omega r}{c}\right)^2}{1 + \left(\frac{\omega r}{c}\right)^2} 4\pi r^2 |\underline{v}|^2$$

$$= \frac{1}{2} \varrho c \frac{\left(\frac{\omega}{c}\right)^2}{4\pi \left[1 + \left(\frac{\omega r}{c}\right)^2\right]} \underbrace{(4\pi r^2 |\underline{v}|)^2}_{\text{volume velocity},\, \underline{q}}. \tag{9.10}$$

In the near field we have $2\pi r/\lambda = \omega r/c \ll 1$, allowing us to write

$$\overline{P} = \frac{1}{2} \varrho c \frac{\left(\frac{\omega}{c}\right)^2}{4\pi} (4\pi r^2 |\underline{v}|)^2 = \frac{1}{2} \frac{\varrho \omega^2}{4\pi c} |\underline{q}_0|^2. \tag{9.11}$$

With $|v| \sim 1/r^2$, the following also holds,

$$(4\pi r^2 |\underline{v}|)^2 \approx |\underline{q}_0|^2 = \text{const}, \tag{9.12}$$

which actually means in the near field that the volume velocity, $\underline{q}$, is fairly independent of the distance, r, and converges to $\underline{q}_0$. This primary volume velocity, $\underline{q}_0$, is called the *source strength* of spherical radiators.

The active power transmitted, $\overline{P}$, does not depend on the distance r, given that the medium is lossless. As a result of this and the fact that active power flows through all spherical shells, we can write

$$\overline{P} = 4\pi r^2 \, \text{Re}\{\underline{I}\} \neq f(r). \tag{9.13}$$

The term 4π in the denominator of equation (9.11) denotes the full spherical angle, $\Omega_\Sigma = 4\pi$. If a 0th-order spherical source with the source strength $\underline{q}_0$ radiates into a smaller spherical angle, Ω_1, that is only a section of the available volume, then the radiated power increases by a ratio of $4\pi/\Omega_1$. Since this power is only radiated into the smaller angle the intensity, $\text{Re}\{\underline{I}\}$, in this section increases by $(4\pi/\Omega_1)^2$ or $20 \lg(4\pi/\Omega_1)$ dB.

This relationship is of practical relevance, for instance, for horn loudspeakers, further for all 0th-order spherical sound sources when placed in front of a wall or in a corner or edge of a room. The following level increases result from such placements[2],

- Placement in front of a wall (hemisphere) $\Longrightarrow$ +6 dB
- Placement in a room edge (quarter sphere) $\Longrightarrow$ +12 dB
- Placement in a corner ($1/8^{\text{th}}$ sphere) $\Longrightarrow$ +18 dB

Point Sources of 0th Order (Monopoles)

In the 0th-order spherical sound field we have

$$\frac{p(r)}{\underline{v}(r)} = \frac{1}{\frac{1}{\varrho c} + \frac{1}{j\omega \varrho r}}, \quad \text{or} \quad \underline{g}_\rightarrow \frac{e^{-j\beta r}}{r} = \frac{\underline{v}(r)}{\frac{1}{\varrho c} + \frac{1}{j\omega \varrho r}}, \tag{9.14}$$

from which follows

$$\underline{g}_\rightarrow = \frac{4\pi \, \underline{v}(r) \, r^2}{4\pi \left(\frac{r}{\varrho c} + \frac{1}{j\omega \varrho} \right)} e^{+j\beta r}. \tag{9.15}$$

We now let the radius of the sphere go to zero while keeping $\underline{g}_\rightarrow$ constant. In this way we obtain

$$\lim_{r \to 0} [4\pi r^2 \, \underline{v}(r)] = \underline{q}_0, \tag{9.16}$$

---

[2] Please note that loudspeakers in closed cabinets become spherical radiators at low frequencies – refer to Section 9.5. Their low-frequency-response can thus be optimized by appropriate placement

from which follows

$$\lim_{r \to 0} \underline{g}_{\to} = \frac{j\omega \varrho \, \underline{q}_0}{4\pi} . \tag{9.17}$$

Finally we arrive at the the sound field of the *point source* of 0th-order, which is also known as *monopole*,

$$\underline{p}_{\to}(r) = \frac{j\omega \varrho}{4\pi} \underline{q}_0 \frac{e^{-j\beta r}}{r} . \tag{9.18}$$

Any 0th-order spherical sound source, that is, any breathing sphere, can be represented by an equivalent monopole with the same source strength, $UL(q)_0$.

9.2 Spherical Sound Sources of 1st Order

A rigid sphere may oscillate according to the sketch in Fig. 9.2 (b), and a sound field created in this way is called a 1st-order spherical sound field. Such a sound field is no longer point-symmetric, which means that the shells around the sphere do not represent areas of equal phase. This may also be expressed as $\partial/\partial\delta \neq 0$ and $\partial/\partial\varphi \neq 0$.

Since the problem is axial-symmetric, it is sufficient to deal with one section through the sphere, and in this case we will take a vertical section along the x-axis. The following boundary condition is valid for the radial component on the surface of the sphere,

$$\underline{v}(\delta) = \underline{v}(0) \cos \delta . \tag{9.19}$$

It will be shown in the following section that the fields of two complementary monopoles with opposite phase can be combined to create a sound field as of an oscillating sphere. The solution of the wave equation for the 1st-order spherical sound field can be derived relatively easily by exploiting this fact.

Point Sources of 1st Order (Dipoles)

Two point sources with equal strength but of opposite phase, that is $\underline{q}_1 = -\underline{q}_0$ and $\underline{q}_2 = +\underline{q}_0$, are positioned a distance $2d$ apart, forming a so-called *dipole*. Due to the linearity of the wave equation, the sound field of this arrangement is given by superposition of the two individual sound fields, namely,

$$\underline{p}_{\to} = \frac{j\omega \varrho}{4\pi} \underline{q}_0 \left(\frac{e^{-j\beta r_2}}{r_2} - \frac{e^{-j\beta r_1}}{r_1} \right) . \tag{9.20}$$

Figure 9.3 (a) illustrates this situation. Since the two 0th-order point sources have zero radius, possible reflection or diffraction caused by their presence need not be considered.

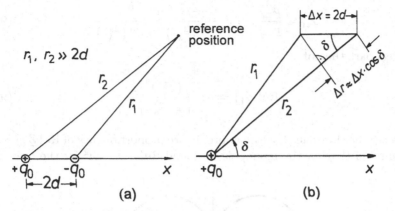

Fig. 9.3. Derivation of the dipole sound field. **(a)** two monopoles of opposite phase at a distance, $2d$, **(b)** equivalent with only one monopole

The next step is to perform a limes-operation in such away that $2d$ goes to zero, while, by definition, the *dipole momentum*, $\underline{\mu}_d = 2d\underline{q}_0$, is kept constant. This condition prohibits the two monopoles from canceling each other, and we can write

$$\underline{p}_\to = \frac{j\omega\varrho}{4\pi}\underline{\mu}_d \lim_{2d\to 0}\left[\frac{1}{2d}\left(\frac{e^{-j\beta r_2}}{r_2} - \frac{e^{-j\beta r_1}}{r_1}\right)\right]. \tag{9.21}$$

Figure 9.2 **(b)** illustrates that the previous equation can also be interpreted as the result of the differentiation of a monopole sound field in the x-direction. By taking $\partial x = \partial r/\cos\delta$ we get

$$\frac{\partial}{\partial x}f(x) = \lim_{\Delta x\to 0}\left[\frac{f(x+\Delta x) - f(x)}{\Delta x}\right], \quad \text{where} \quad \Delta x = 2d. \tag{9.22}$$

The resulting solutions of the wave equation for the dipole field are as follows – outward-progressing waves only,

$$\underline{p}_\to(r,\delta) = \frac{j\omega\varrho}{4\pi}\underline{\mu}_d \cos\delta\,\frac{\partial}{\partial r}\left(\frac{e^{-j\beta r}}{r}\right)$$

$$= \frac{-j\omega\varrho}{4\pi}\underline{\mu}_d \cos\delta\left(\frac{1}{r^2} + \frac{j\beta}{r}\right)e^{-j\beta r}, \tag{9.23}$$

$$\underline{v}_\to(r,\delta) = \frac{\underline{\mu}_d}{4\pi}\cos\delta\left(-\frac{2}{r^3} - \frac{2j\beta}{r^2} + \frac{\beta^2}{r}\right)e^{-j\beta r}. \tag{9.24}$$

The solution for $\underline{v}$ has again been derived via *Euler*'s equation. Please note that the sound pressure possesses a $1/r^2$-component, which means that it has a near field. The field impedance is, with $\beta = 2\pi/\lambda = \omega/c$,

$$\underline{Z}_f = \frac{\underline{p}_\rightarrow}{\underline{v}_\rightarrow} = \varrho c \; \frac{-j\frac{2\pi r}{\lambda} + \left(\frac{2\pi r}{\lambda}\right)^2}{2 - \left(\frac{2\pi r}{\lambda}\right)^2 + j2\left(\frac{2\pi r}{\lambda}\right)} \; . \tag{9.25}$$

The real part thereof is

$$\mathrm{Re}\{\underline{Z}_f\} = \varrho c \; \frac{\left(\frac{2\pi r}{\lambda}\right)^4}{4 + \left(\frac{2\pi r}{\lambda}\right)^4} \; . \tag{9.26}$$

In the near field we thus have an approximate proportionality of $\mathrm{Re}\{\underline{Z}_f\} \sim \omega^4$. Please recall that for the monopole we found $\mathrm{Re}\{\underline{Z}_f\} \sim \omega^2$ for the near field.

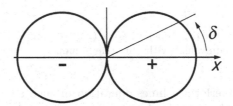

Fig. 9.4. Directional characteristics, Γ, of a dipole sound source

The dipole sound field shows a directional characteristic for both sound pressure and particle velocity,

$$\Gamma = \frac{\underline{p}_\rightarrow(r,\delta)}{\underline{p}_\rightarrow(r,0)} = \frac{\underline{v}_\rightarrow(r,\delta)}{\underline{v}_\rightarrow(r,0)} = \cos\delta \, , \tag{9.27}$$

as plotted in Fig. 9.4. Note that the plot only shows the vertical plane, but the directional characteristics are axial-symmetric around the x-axis.

We see a figure-of-eight characteristic that complies with the boundary conditions of the rigid oscillating sphere. As a result, the sound field of oscillating spheres can be represented by 1st-order spherical point sources (dipoles).

9.3 Higher-Order Spherical Sound Sources

Any sound-field can be considered to be composed of a series of orthogonal spherical harmonics of different orders. These spherical harmonic waves are eigen-functions of the wave equation in spherical coordinates. The first two, the spherical waves of 0th and 1st order, have been introduced in the preceding sections. Spherical waves of higher order are radiated by spheres with surfaces that oscillate with velocities determined by higher-order spherical functions. Figure 9.2 (c) depicts one possible 2nd-order spherical vibration. Spherical sources of higher order can be represented by combinations of monopoles and dipoles.

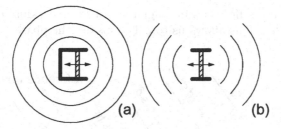

Fig. 9.5. Schematic plots of the sound fields of *breathing*, (**a**), and *oscillating* sound sources, (**b**)

For spherical sound emitters of nth order in the near field, the resistive part of the field impedance, $\text{Re}\{\underline{Z}_f\}$, increases with frequency as follows.

$$\text{Re}\{\underline{Z}_f\} \sim \omega^{2(n+1)} . \tag{9.28}$$

If the linear dimensions of the emitter are small compared to wavelength, that is $2\pi r_0 \ll \lambda$, the radiation of higher-order spherical waves may be neglected at low frequencies, making it possible to approximate most sound sources at low frequencies with either a monopole or a dipole. In fact, the monopole or 0th-order spherical source provides good low-frequency approximation for all *breathing* sound sources such as a loudspeaker mounted in a cabinet, while the dipole or 1st-order spherical source serves well to approximate *oscillating* sources such as a loudspeaker without a baffle– see Fig. 9.5 (a, b).

9.4 Line Arrays of Monopoles

Arrangements of several sources along a line in space are called *line arrays* (... linear arrays) and play an important role in practical applications. Sharply bundled radiation can be achieved with these sources because the sound fields of the individual sources interfere with each other. A common application of this principle is the line array of loudspeakers.

In the following discussion, the directional characteristics of linear arrays composed of monopoles will be considered by restricting ourselves to the sound field far away from the array.

Line Array of Identical and Equidistant Monopoles

In an arrangement like the one depicted in Fig. 9.6, we look at a reference point at a distance of $r_0 \gg 2h$. The sum of the contributions of all monopoles of the array is

$$\underline{p}_{\rightarrow}(r = r_0, \delta) = \frac{j\omega \varrho \, \underline{q}_0}{4\pi} \sum_{i=1}^{n} \frac{e^{-j\beta \, [r_0 - (i-1)2d \cos \delta]}}{r_0 - (i-1)2d \cos \delta} . \tag{9.29}$$

We may neglect the differences in magnitude of the individual contributions because of $r_0 \gg 2h$. This allows us to only consider the phase differences.

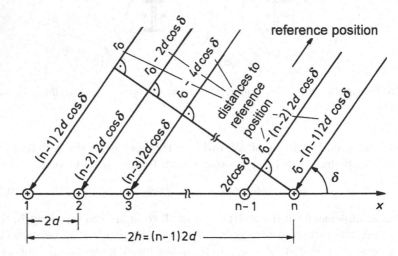

Fig. 9.6. Linear array with monopole sources

This leads to the following approximation,

$$\underline{p}_\rightarrow(r,\delta) = \frac{j\omega\,\varrho\,\underline{q}_0}{4\pi}\,\frac{e^{-j\beta\,r_0}}{r_0}\sum_{i=1}^{n}e^{+j\beta\,(i-1)\,2d\,\cos\delta}\,.\tag{9.30}$$

Substituting $\beta\,d\,\cos\delta$ with b in the expression for the sum, we obtain an expression with a known series summation,

$$\sum e^{+j(i-1)2b} = 1 + e^{+j2b} + e^{+j4b} + \cdots e^{+j2(n-1)b} = \frac{1-e^{+j2nb}}{1-e^{+j2b}}\,.\tag{9.31}$$

Writing with an expansion using $\sin x = (e^{+jx} - e^{-jx})/2j$, yields,

$$\sum e^{+j(i-1)2b} = \frac{e^{+jnb}}{e^{+jb}}\left[\frac{e^{-jnb}-e^{jnb}}{e^{-jb}-e^{jb}}\right] = e^{+j(n-1)b}\,\frac{\sin(nb)}{\sin(b)}\,.\tag{9.32}$$

The term $\sin(nb)/\sin b$ determines the directional characteristic. We shall discuss it more easily in the following paragraph, where a continuously loaded line of monopoles is dealt with.

Continuously Loaded Line Array

First we perform a limit operation by letting the distance between the individual monopoles, $2d$, and, consequently, b, go to zero. With the length of line array, $2h$, kept constant, we then get $n \to \infty$. To also keep the total source strength, $n \underline{q}_0 = \underline{q}'(x)\,2h$, constant, we normalize by n, with $\underline{q}'(x)$ being a constant velocity load. The result of this operation, namely,

$$\lim_{2d \to 0} \frac{\sin(nb)}{n \sin(b)} = \frac{\sin(nb)}{nb} = \mathrm{si}(nb),\qquad(9.33)$$

is a so-called *sinc-* or *si-function* (sinus cardinalis). With $2h \approx n\,2d$ and, therefore, $nb = \beta h \cos \delta$, it follows that

$$\Gamma = \mathrm{si}\left(\beta h \cos \delta\right).\qquad(9.34)$$

This is the directional characteristic of the in-phase, continuously loaded line array with a constant source-strength load. The formula can also be loosely applied to arrays with a limited number of monopoles.

As an example, Figure 9.7 illustrates a line array with a length of two wavelengths, $2h = 2\lambda$, and $\beta h = 2\pi$. The upper panel illustrates the directional characteristics in *Cartesian* coordinates. The lower panel shows the typical club-shaped form of the beam in spherical coordinates – vertical section only.

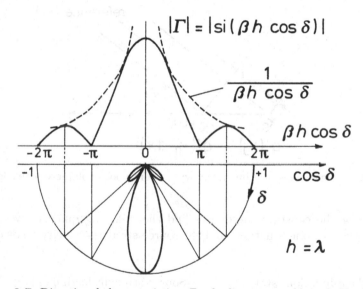

Fig. 9.7. Directional characteristics, Γ, of a line array of length $2h = 2\lambda$

9.5 Analogy to *Fourier* Transforms as Used in Signal Theory

In the preceding section we assumed a continuous source-strength load, $q'(x)$, having the dimension [volume velocity/length]. We continued to presume that each point on the line acts as a monopole – shown in Fig. 9.8. With the following integration we can now calculate the sound pressure at the observation point, $r_0 \gg 2h$. The expression in front of the integral is a constant term for a given r_0.

$$\underline{p}_{\rightarrow}(r, \delta) = \underbrace{\frac{j\omega\,\varrho}{4\pi}\,\frac{e^{-j\beta\,r_0}}{r_0}}_{\text{const.}}\,\int_{-h}^{+h}\underline{q}'(x)\,e^{+j(\beta\cos\delta)x}\,dx\,. \tag{9.35}$$

This expression is clearly isomorphic to the well-known *Fourier* integral from signal theory, which can be written as

$$\underline{S}(\omega) = \int s(t)\,e^{-j\omega\,t}dt\,, \quad \text{or, symbolically, as} \tag{9.36}$$

$$s(t) \;\circ\!\!-\!\!\bullet\; \underline{S}(\omega)\,, \tag{9.37}$$

where t corresponds to x and ω corresponds to $-\beta\cos\delta$.

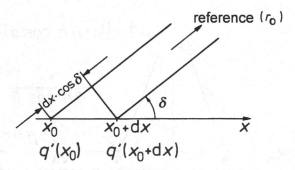

Fig. 9.8. Sound fields of a line array with a continuous volume-velocity load

Disregarding the constant factor, we find the following quantitative analogies between the time functions and the source-strength-velocity loads of line arrays.

Time function, $s(t)$		Source-strenght load, $q'(x)$
$\circ$		$\circ$
$\bullet$	$\Longleftrightarrow$	$\bullet$
Spectrum, $|\underline{S}(\omega)|$		Directional characteristic, $\Gamma(\delta)$

The correspondences shown in Table 9.1 lists examples we have treated so far[3].

---

[3] For the definition of Γ see (9.27). In the table, the directional characteristics, $\Gamma(\delta)$, have been normalized so that their maxima equal one. $\vartheta(z)$ is called *Dirac impulse*. It is a special mathematical *distribution* that picks out the value of a

Table 9.1. Some examples of the equivalence of time signal and spectrum v.s. velocity load and directional characteristics

Linear array with constant source-strength load	Rectangular impulse
$\begin{cases} \underline{q}' = \text{const} & \text{for } -h < x < +h \\ \underline{q}' = 0 & \text{others} \end{cases}$	$\begin{cases} s(t) = \text{const} & \text{for } -\tau < t < \tau \\ s(t) = 0 & \text{others} \end{cases}$
$\circ\!\!\!\!-\!\!\!\bullet$	$\circ\!\!\!\!-\!\!\!\bullet$
$\Gamma = \text{si}\,(-h\,\beta\,\cos\delta)$	$S(\omega) = 2\tau\,\text{si}\,(\tau\,\omega)$
Monopole	***Dirac* impulse**
$\underline{q}'(x) = \underline{q}_0\,\vartheta(x)$	$s(t) = \vartheta(t)$
$\circ\!\!\!\!-\!\!\!\bullet$	$\circ\!\!\!\!-\!\!\!\bullet$
$\Gamma = 1$	$S(\omega) = 1$
Dipole	**Double *Dirac* impulse**
$\underline{q}'(x) = \underline{\mu}_d\,\frac{\mathrm{d}}{\mathrm{d}x}\vartheta(x) = \underline{\mu}_d\,\vartheta'(x)$	$s(t) = \frac{\mathrm{d}}{\mathrm{d}t}\vartheta(t) = \vartheta'(t)$
$\circ\!\!\!\!-\!\!\!\bullet$	$\circ\!\!\!\!-\!\!\!\bullet$
$\Gamma = \cos\delta$	$S(\omega) = \omega$

At the end of this section we would like to discuss two additional directional characteristics which are relevant from an application point of view and can also be obtained from analogous relationships in signal theory.

• A source-strength load with a *Gaussian* envelope leads to a *Gaussian* directional characteristic. This is a beam without side lobes – shown in Fig. 9.9 (a)

• A source-strength load with a phase shift increasing linearly with position,

$$q'_2(x) = q'_1(x)\,\mathrm{e}^{-\mathrm{j}\beta\cos(\Delta\delta)\,x} \quad\circ\!\!\!\!-\!\!\!\bullet\quad \Gamma_2 = \Gamma_1\,[\beta\,\cos\delta - \beta\cos(\Delta\delta)] \quad (9.38)$$

leads to a tilted directional characteristics – schematically shown in Fig. 9.9 (b). The *shifting theorem* of *Fourier* transforms has been used here[4].

---

function at the position of its argument as follows, $\int_{-\infty}^{+\infty} y(z)\,\vartheta(z-z_0)\,\mathrm{d}z = y(z_0)$. The area under the *Dirac* impulse is $\int_{-\infty}^{\infty}\vartheta(z)\,\mathrm{d}z = 1$. NB: *Dirac* impulses are usually written $\delta(z)$ instead of $\vartheta(z)$.

[4] An application of this algorithm, based on the directional equivalence of emitters and receivers – see next Section – is the electronic steering of SONAR antennas

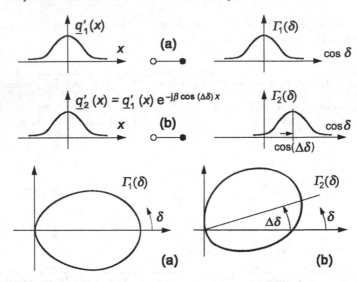

Fig. 9.9. (a) Shaping, and (b) shifting of directional characteristics (schematic)

9.6 Directional Equivalence of Sound Emitters and Receivers

When a reversible transducer or transducer array is operated as sound emitter, its directional characteristic is equivalent to its directional sensitivity characteristic when operated as a receiver.

This can be shown by using the following two elements,

M ... a transducer with arbitrary directional characteristics
X ... an auxiliary point source with monopole characteristics

The point source is positioned far away from the transducer under consideration. The proof is done in two steps as follows.

- The transducer is fed with an electric current, $\underline{i}_0$. At the position of the auxiliary source we then have

$$|\underline{p}_X| = |\underline{T}_{ip}(\omega, \phi, \delta, r)| \, |\underline{i}_0| \tag{9.39}$$

- The auxiliary point source emits a volume velocity amounting to its source strength, q_0. At the position of the transducer, which is not present at this point, we get

$$|\underline{p}_M| = \frac{\omega \varrho}{4\pi r} |\underline{q}_0| \tag{9.40}$$

If the transducer is now introduced into the sound field, a voltage, $\underline{u}_1$, can be measured at its electric output port according to

$$|\underline{u}_1| = |\underline{T}_{\text{pu}}(\omega, r, \delta, \phi)| \, |\underline{p}_M|. \tag{9.41}$$

Since the sound field is linear and passive, reversibility according to (4.16) applies as follows,

$$\left. \left| \frac{\underline{i}_0}{\underline{p}_X} \right| \right|_{q=0} = \left. \left| \frac{\underline{q}_0}{\underline{u}_1} \right| \right|_{i=0}, \tag{9.42}$$

and, hence,

$$\left| \frac{T_{\text{ip}}}{T_{\text{pu}}} \right| = \left| \frac{\omega \varrho}{4\pi r} \right|. \tag{9.43}$$

The first thing that can be seen from this equation is that the transfer coefficient of the transmitter in emitting function increases with frequency with respect to the transfer coefficient for receiver operation (sensitivity). This actually means that

Transducers receive low frequencies better than they emit them!

A good example for this law is a small reversible microphone.

Finally, to show that the directional characteristic, Γ, for transmitter operation is identical to the characteristic for receiver operation, we set

$$\frac{|\underline{T}_{\text{ip}}(\phi, \delta)|}{|\underline{T}_{\text{pu}}(\phi, \delta)|} = \frac{|\underline{T}_{\text{ip}}(0,0)| \, \Gamma_{\text{ip}}}{|\underline{T}_{\text{pu}}(0,0)| \, \Gamma_{\text{pu}}} = \left| \frac{\underline{T}_{\text{ip}}(0,0)}{\underline{T}_{\text{pu}}(0,0)} \right|, \tag{9.44}$$

which results in

$$\Gamma_{\text{ip}} = \Gamma_{\text{pu}}. \tag{9.45}$$

In the context of the examples that we have dealt with in this book, we find the following correspondences,

Pressure receiver	$\Longleftrightarrow$	0th-order spherical source
Pressure-gradient receiver	$\Longleftrightarrow$	1st-order spherical source
Line microphone	$\Longleftrightarrow$	line array with constant source-strenght load and $90°$ shifted directional characteristic

10

Piston Membranes, Diffraction and Scattering

In the last chapter the sound field produced by point-source arrangements was calculated by linear superposition of individual spherical sound fields. For continuously loaded line arrays we further substituted the source strength of the individual sources with a length-specific source-strength load, $\underline{q}'_0(x)$. The source-strength load has the dimension [volume-velocity/length].

It seems natural to extend this method to area radiators like oscillating membranes. There we will obtain an area-specific source-strength load, $\underline{q}''_0(x, y) = \underline{v}(x, y)$, with the dimension [volume velocity/area], which is indeed equal to [particle velocity].

In the case of a line array composed from point sources, reflection and diffraction of the sound field due to the array itself can be disregarded. This does no longer hold for area radiators because the area can act as both reflector and diffractor. Computations related to the sound fields of such radiators can thus become complicated[1].

There is a particular case, however, where reflection and diffraction do not occur. In this book, we will restrict ourselves to just this case, which is a flat membrane in an infinitely extended, rigid plane baffle. The *Huygen's* principle can be applied to this special case in an elementary way by simply superimposing 0th-order point sources[2]. In addition, many practical problems can be approximated by this special case.

---

[1] The sound-field can, for example, be calculated with the *Kirchhoff–Helmholtz* integral equation that determines the sound field inside an enclosed space from the sound-pressure and the pressure-gradient distributions on an enclosing surface

[2] The *Kirchhoff–Helmholtz* integral equation then reduces to the so-called *Rayleigh* or *Huygens–Helmholtz* integral – see Section 10.1

10.1 The *Rayleigh* Integral

A small vibrating piston results in a 1st-order spherical sound field– see Fig. 9.5 (b). Yet, building this piston into an infinitely extended, plane and rigid baffle prohibits hydrodynamic shorting between the front and back of the baffle. This results in a hemispherical sound field of 0th order radiating into half of the space – shown in Fig. 10.1. The hemispherical sound field is

$$\underline{p}_{\rightarrow}(r) = \frac{j\omega\,\varrho}{2\pi}\,\underline{q}_0\,\frac{e^{-j\beta r}}{r}\,, \tag{10.1}$$

where 2π is the spatial angle of a hemisphere. Assuming that the baffle is flat, this sound field has no normal components in the plane of the baffle and, therefore, the baffle cannot cause any reflection or diffraction.

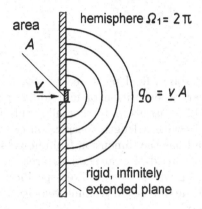

Fig. 10.1. Hemispherical sound field, originating from a point source in a flat and rigid baffle

Now consider an oscillating membrane with an area, A_0, in this infinitely extended, flat and plane baffle where all area elements oscillate perpendicularly to the area. This arrangement can be considered to be the superposition of an infinite number of adjacent monopoles, $d\underline{q}_0 = \underline{v}\,dA$ – shown in Fig. 10.2. The total sound pressure at an observation point is found by the following superposition of these monopoles,

$$\underline{p}_{\rightarrow}(r) = \frac{j\omega\,\varrho}{2\pi}\int_{A_0}\underline{v}\,\frac{e^{-j\beta r}}{r}\,dA\,. \tag{10.2}$$

In acoustics, this integral is usually called *Rayleigh integral*. It is valid at all distances from the membrane.

It is worth noting that $\underline{v}\,(x,y)$ does not need to be the same everywhere on the membrane, that is, its value may depend on the position of the membrane.

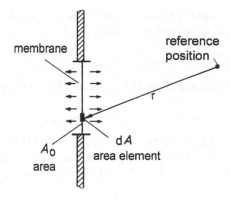

Fig. 10.2. Sound field in front of a membrane in a flat and rigid baffle

For the *Rayleigh* integral to converge without additional assumptions, such as propagation losses in the medium, it is necessary that A_0 is finite. Further, propagation in the considered hemisphere must be free of obstacles.

10.2 *Fraunhofer's* Approximation

Fraunhofer's approximation applies when the distance from the reference point to the membrane is very large in comparison to the linear dimensions of the radiating membrane. Figure 10.3 depicts the situation to be discussed.

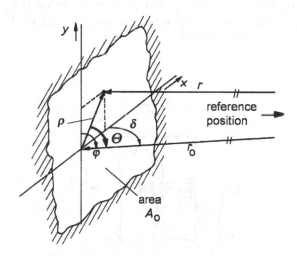

Fig. 10.3. *Fraunhofer's* approximation

The quantity r_0 is the distance from the reference point to any position on the membrane with the area A_0. That position is preferably the membrane's center of gravity. The angle between r and r_0 is very small, and the two lines are practically parallel. The following linear approximation is thus applicable as a result,

$$r \approx r_0 - \rho \cos \Theta,$$ (10.3)

where Θ is the angle between r_0 and the ρ-axis. We can write this in *Cartesian* coordinates as

$$r \approx r_0 - x \cos \delta - y \cos \varphi,$$ (10.4)

where δ is the angle between r_0 and the x-axis, and φ is the angle between r_0 and the y-axis. The factor $1/r$ can brought out in front of the integral because $1/r \approx 1/r_0$ – as we saw in Section 9.5 – which brings us to

$$\underline{p}_{\rightarrow}(r, \delta, \varphi) = \frac{\mathrm{j}\omega\,\varrho}{2\pi\,r_0}\,\mathrm{e}^{-\mathrm{j}\beta r_0} \int_{-\infty}^{+\infty} \int_{-\infty}^{+\infty} \underline{v}(x, y)\;\mathrm{e}^{\mathrm{j}(\beta\,\cos\delta)x}\,\mathrm{e}^{\mathrm{j}(\beta\,\cos\varphi)y}\;\mathrm{d}x\,\mathrm{d}y.$$ (10.5)

This expression is again isomorphic to a *Fourier* transform. It is actually a two-dimensional, spatial *Fourier* transform, where $\beta \cos \delta$ is the phase coefficient in x-direction, and $\beta \cos \varphi$ is the phase coefficient in y-direction. Recall that $\beta = 2\pi/\lambda$.

10.3 The Far Field of Piston Membranes

This section deals with the sound field produced by rigid membranes, so-called *piston membranes* in infinitely extended, rigid, flat baffles. In other words, we speak of baffled pistons with identic $\underline{v}(x, y)$ everywhere on the piston. Such piston membranes can serve as models for loudspeakers in large baffles as long as the membranes of the loudspeakers vibrate in phase.

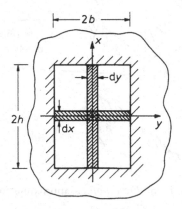

Fig. 10.4. Rectangular membrane

Rectangular Piston Membranes

The sound field of a rectangular piston membrane in an infinitely extended, rigid, flat baffle can be computed with *Fraunhofer*'s approximation according to Fig. 10.4. By symbolizing the constant factors left of the integral as $\underline{C}_1$ or $\underline{C}_2$, respectively, we can write *Fraunhofer*'s approximation and its *Fourier* transform as

$$\underline{p}_{\rightarrow}(r, \delta, \varphi) = \underline{C}_1 \int_{-b}^{+b} \left[\int_{-h}^{+h} \underline{v}\, e^{j(\beta \cos \delta)x}\, e^{j(\beta \cos \varphi)y}\, dx \right] dy \quad (10.6)$$

$$= \underline{C}_2 \underbrace{\int_{-h}^{+h} e^{j(\beta \cos \delta)x}\, dx} \underbrace{\int_{-b}^{+b} e^{j(\beta \cos \varphi)y}\, dy}, \quad (10.7)$$

$$\underline{p}_{\rightarrow}(r, \delta, \varphi) \circ\!\!-\!\!\bullet\, \underline{C}_2\, 2h\, \underbrace{\text{si}\,(h\,\beta \cos \delta)}_{\Gamma(\delta)}\, 2b\, \underbrace{\text{si}\,(b\,\beta \cos \varphi)}_{\Gamma(\varphi)}. \quad (10.8)$$

Because each of the two integrals is actually a one-dimensional *Fourier* integral, the integration can be performed by well known rules. The result is (10.8), where the first term on the right side includes the directional characteristics with respect to δ and the second with respect to φ. The total two-dimensional directional characteristics is formed by multiplying two one-dimensional characteristics. Each represents a continuously loaded line array, one on the x-axis and one on the y-axis, which can be written as

$$\Gamma(\delta, \varphi) = \Gamma(\delta)\, \Gamma(\varphi). \quad (10.9)$$

The third dimension, r, is not considered as the derivation above only deals with the sound field far away from the membrane.

Circular Piston Membranes

The calculation of circular pistons is slightly more complicated and requires a transformation into polar coordinates – illustrated in Fig. 10.5. The result of the calculation is given below,

$$\Gamma(r) = \frac{2\,\mathbf{J}_1(R\,\beta \cos \Theta)}{R\,\beta \cos \Theta}, \quad (10.10)$$

where Θ is the angle between the membrane and a line leading from the observation point to the middle of the membrane. R is the radius of the membrane, and $\mathbf{J}_1$ is the the first order *Bessel* function of the first kind.

The functions $\sin x/x = \text{si}\,(x)$ and $\mathbf{J}_1(x)/x$ look very similar. This means that circular membranes have directional characteristics similar to rectangular ones. The example shown in Fig. 10.6 illustrates the similarities when we choose $b = h = R$.

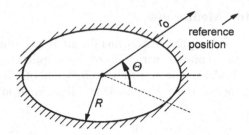

Fig. 10.5. Circular membrane from a perspective view

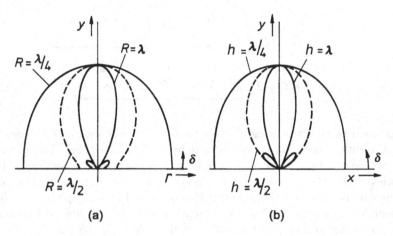

Fig. 10.6. Examples for directional characteristics of (**a**) circular membrane, (**b**) quadratic membrane

10.4 The Near Field of Piston Membranes

Because the sound field close to a membrane can not be computed by *Fraunhofer*'s approximation, the *Rayleigh* integral itself must be solved. Discrete numerical methods are commonly used to accomplish this.

Zone Construction after *Huygens* and *Fresnel*

In this discussion, we introduce a traditional method of solving this integral by graphic interpretation of the conditions near the membrane. This technique is called *Huygens–Fresnel* zone construction and can be applied to piston membranes, or more specifically, vibrating plane areas where v is constant across the area.

As shown in Fig. 10.7, the radiating area is subdivided into ring-shaped zones. The average difference in radial distance of two adjacent zones is $\lambda/2$. As a result, the average contributions rendered by two adjacent zones have a phase difference of 180°. The ring zones are constructed in such a way that

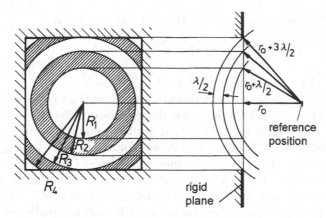

Fig. 10.7. The construction of *Huygens–Fresnel* zones

all complete zones make identical contributions, and adjacent complete zones cancel each other out. The magnitude of the resulting sound pressure at the observation point is estimated from the contributions of the remaining areas after cancelation.

We will now prove that the contributions of the complete zones are actually identical. On the one hand, the area of a ring zone is

$$A_n = \pi (R_n^2 - R_{n-1}^2)\,. \tag{10.11}$$

By applying *Pythagoras'* law repeatedly, we find

$$A_n = \pi \left[\left(r_0 + n\frac{\lambda}{2} \right)^2 - \left(r_0 + \frac{(n-1)\lambda}{2} \right)^2 \right] = \pi\lambda \left[r_0 + (2n-1)\frac{\lambda}{4} \right]\,. \tag{10.12}$$

On the other hand, the average distance of a ring zone from the reference point is

$$\bar{r}_n = \frac{1}{2} \left[r_0 + \frac{n\lambda}{2} + r_0 + (n-1)\frac{\lambda}{2} \right] = \left[r_0 + (2n-1)\frac{\lambda}{4} \right]\,. \tag{10.13}$$

The resulting ratio of the ring-zone area to the average distance is constant and equal to

$$\frac{A_n}{\bar{r}_n} = \pi\lambda\,. \tag{10.14}$$

Now we will calculate the contribution of a single ring zone to the sound pressure at the reference point, namely,

$$\underline{p}_{\rightarrow,\,\mathrm{zone}} = \frac{\mathrm{j}\omega\varrho}{2\pi} \int_{A_1} \underline{v}\,\frac{\mathrm{e}^{-\mathrm{j}\beta r}}{r}\,\mathrm{d}A = \frac{\mathrm{j}\omega\varrho\,2\pi\,\underline{v}}{2\pi} \int_{r_0}^{r_0+\lambda/2} \mathrm{e}^{-\mathrm{j}\beta r}\,\mathrm{d}r\,, \tag{10.15}$$

where $A = \pi(r^2 - r_0^2)$, and $\mathrm{d}A = 2\pi r\,\mathrm{d}r$.

Taking the 1$^\text{st}$ zone as an example, this can be rewritten as

$$\underline{p}_{\rightarrow,\text{zone 1}} = \varrho c\, \underline{v}\, e^{-j\beta r_0} - \varrho c\, \underline{v}\, e^{-j\beta(r_0+\lambda/2)} = \varrho c\, 2\underline{v}\, e^{-j\beta r_0}, \qquad (10.16)$$

where $\beta = 2\pi/\lambda = \omega/c$ and $e^{-j\beta\lambda/2} = -1$.

The following interpretation follows directly from this expression. The first term on the right side of the equation describes an undisturbed plane wave. The second term stands for a wave that results from diffraction at the outer rim of the ring zone. Therefore, we actually find two times the magnitude of the sound pressure magnitude of a plane wave at the reference point. This means, for example, that if a plane wave is interrupted by an infinitely extended, or at least very large baffle with a circular hole in it – see Fig. 10.11 – the sound pressure at the observation point behind the baffle can be up to two times larger!

The following rules are helpful for graphical evaluation of *Huygens–Fresnel* zone construction. The contribution of incomplete zones is considered to be approximately proportional to the ratio of the remaining area to the complete area and must assume the appropriate sign. When summing up, one starts by letting the contribution of the first half zone stand while its second half cancels out with the first half of the adjacent zone, and so on. This procedure allows the resulting field to be interpreted as the sum of a plane wave and interfering diffracted waves from the rims of the radiating areas.

On-Axis Sound Pressure and Radiation Impedance of Circular Piston Membranes

We will now discuss the sound pressure of a diverging wave, $|\underline{p}_{\rightarrow}(r_0)|$, on the middle axis of a circular piston membrane. This is an example of the near field of a circular piston membrane in an infinitely extended, rigid, plane baffle with radius R.

For a given r_0 and wavelength, λ, the number of *Huygens–Fresnel* zones is equal to

$$\nu_\text{z} = \frac{\sqrt{R^2 + r_0^2} - r_0}{\lambda/2}. \qquad (10.17)$$

Maxima and minima occur whenever ν_z is exactly an integer number, which occurs for

$$r_0 = \frac{\left(\frac{R}{\lambda}\right)^2 - \left(\frac{\nu_\text{z}}{2}\right)^2}{\nu_\text{z}/\lambda} \doteq \nu_{\text{z, integer}}, \qquad (10.18)$$

where odd integers result in maxima and even integers result in minima.

Exactly one zone exists for $\nu_\text{z} = 1$. For distances greater than $r_0|_{\nu_\text{z}=1}$, the sound pressure decreases monotonically, which means that we then are in the far field. The equation cannot be fulfilled for $R < \lambda$, meaning that there is no zero. There is a finite number of zeros for $R > \lambda$ – depicted in Fig. 10.8.

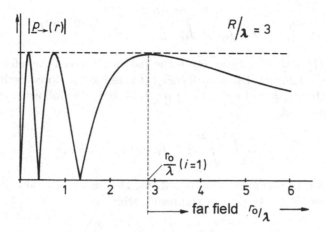

Fig. 10.8. Sound pressure on the axis of a circular piston membrane as a function of distance and wavelength

Calculation of the acoustic power transmitted by a piston in a baffle requires the real component of the radiation impedance, called the radiation resistance, that is

$$r_{\text{rad}} = \text{Re}\{\underline{Z}_{\text{rad}}\}, \quad \text{with} \quad \underline{Z}_{\text{rad}} = \frac{F}{\underline{v}} = \frac{\int_{A_0} \underline{p}\, dA}{\underline{v}}. \tag{10.19}$$

Now we shall describe the steps of how to calculate this resistance. Details are available in the relevant literature. The calculation is performed using the sound pressure directly on the surface area, A_0, of the radiating piston. Integration across this area is shown in Fig. 10.9.

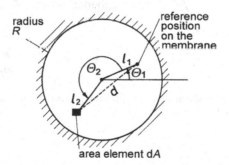

Fig. 10.9. Coordinates for calculating the radiation resistance of a circular piston

For a circular piston membrane the sound pressure follows from the *Rayleigh* integral as follows,

$$\underline{p}_{\rightarrow}(l_1, \Theta_1) = \frac{j\omega \varrho}{2\pi} \int_{l_2=0}^{l_2=R} \int_{\Theta_2=0}^{\Theta_2=2\pi} \underline{v} \, \frac{e^{-j\beta d}}{d} \, l_2 \, dl_2 \, d\Theta_2 , \qquad (10.20)$$

where $d = \sqrt{l_1^2 + l_2^2 - 2 l_1 l_2 \cos \Theta_2}$ is the distance between the reference point and the area element, $dA = l_2 \, dl_2 \, d\Theta_2$. To derive the force acting on the membrane, that is, $\underline{F} = \int \underline{p}_{\rightarrow} \, dA = \int \underline{p}_{\rightarrow} \, l_1 \, dl_1 \, d\Theta_1$, the following integration must be performed,

$$\underline{F} = \int_0^R \int_0^{2\pi} \underline{p}_{\rightarrow}(l_1, \Theta_1) \, l_1 \, dl_1 \, d\Theta_1 . \qquad (10.21)$$

The evaluation of this expression is tedious. Details are available in the relevant literature. We only present the result, which is

$$r_{\text{rad}} = \text{Re}\{\underline{Z}_{\text{rad}}\} = \left(1 - \frac{\mathbf{J}_1(2\beta R)}{\beta R}\right) A_0 \varrho c, \qquad (10.22)$$

where $\mathbf{J}_1$ is the first-order *Bessel* function of the first kind. Figure 10.10 illustrates the results for the circular piston and its area-equivalent 0th-order spherical source – shown in Fig. 8.6 (b). Their courses are clearly related, with the exception of some overshoots that can be explained by diffraction happening at the rim of the circular membrane. The results for rectangular pistons are quite similar.

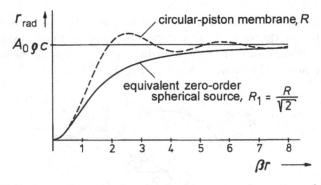

Fig. 10.10. Radiation resistance of a circular-piston membrane as a function of distance and wavelength

10.5 General Remarks on Diffraction and Scattering

The calculations performed above for pistons can be expanded to the phenomenon of *diffraction* caused by circular holes in baffles – see Fig. 10.11. We assume that a plane wave hits such a baffle perpendicularly, and that there is a constant velocity, $\underline{v}(x,y)$, normal to the plane of the area of the hole. The situation is actually identical to the piston in a baffle.

The sound field behind the baffle is calculated with the *Rayleigh* integral, using the method of superposition of 0th-order point-source waves, keeping in mind their phases and mutual interferences.

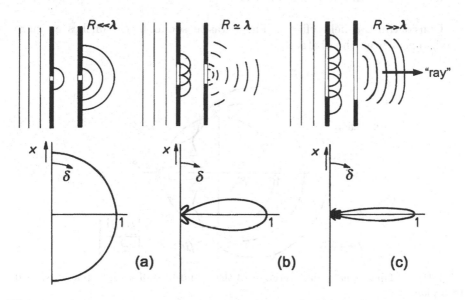

Fig. 10.11. Upper panel: wave-field plots for a circular hole in a baffle when hit by a plane wave. **Lower panel**: corresponding directional characteristics. Linear dimensions of the hole: **(a)** small compared to the wavelength, **(b)** on the order of the wavelength, **(c)** large compared to the wavelength

Figure 10.11 illustrates three typical cases, namely $R \ll \lambda$, $R \approx \lambda$, and $R \gg \lambda$. The upper panels represent the sound fields yielded by the *Rayleigh* integral, the lower panels show their respective directional characteristics, Γ. For $R \ll \lambda$, diffraction causes the waves to propagate into a full hemisphere behind the baffle without any shadowing effects. For $R \approx \lambda$, both diffraction and shadowing occur. We see pronounced side lobes in addition to the main beam. For $R \gg \lambda$, the waves propagate mainly in the direction of the main beam. Side lobes are present but negligible, resulting in a well pronounced beam. The sound wave thus propagates straight along the axis of the beam like a *ray*.

Diffraction also occurs when obstacles of finite dimension are positioned in a sound field. If a sound wave hits such an obstacle, it causes parasitic waves that superimpose themselves upon the original ones. Solving the wave equation consists of the following tasks.

- Formulating the boundary conditions at the surface. For resting, rigid obstacles for example, the normal component of the volume velocity, $v_\perp$, must be zero at the surface

- Summing the original and parasite waves. This results in a sound field that meets the given boundary conditions

- Considering far-field effects. The parasite sound field vanishes at large distances from the obstacle

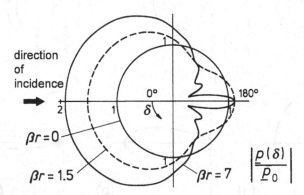

Fig. 10.12. Directional characteristics of the sound field of a rigid sphere exposed to a plane wave

As an example, Figure 10.12 illustrates the sound field resulting from the impingement of a plane wave on a rigid sphere. The figure contains data for three different relative sphere sizes, namely, $\beta R = 2\pi(R/\lambda) = 0$, 1.5, and 7. The curves depict the magnitude ratio $|\underline{p}(\delta)/\underline{p}_0|$, of the sound pressure on the surface of the sphere, p, with respect to the sound pressure in the undisturbed free field, $\underline{p}_0$. The following interesting details become obvious.

On the side facing the incoming wave the sound accumulates (piles up) whereby the sound pressure increases by a factor of up to 2 (6 dB) at high frequencies. One situation where this effect must be taken into account is in the designing of microphones. On the opposite side of the obstacle, the sound pressure is nonzero. There is actually a pronounced maximum at $\delta = 180°$, called a *bright spot*. This effect is caused by waves *creeping* around the sphere and adding in phase at $\delta = 180°$.

The conclusion we draw from this example is that there are two important effects at work. The first is called *scattering* and is seen when the sound waves are bounced back by the surface facing the incoming wave. The other is called *diffraction* and leads to the interference field building up on the opposite side of the obstacle as a result of wave components being deflected into the space behind the obstacle.

11

Dissipation, Reflection, Refraction, and Absorption

The wave equation as derived in Section 7.1 and used so far in this book, is valid for sound propagation in lossless media. The *Helmholtz* form of this equation is

$$\frac{\partial^2 \underline{p}}{\partial x^2} - (j\beta)^2 \underline{p} = 0. \tag{11.1}$$

Its solution for the forward-progressing plane wave can be expressed as

$$\underline{p}_{\rightarrow}(x) = \underline{p}_{\rightarrow} e^{-j\beta x}. \tag{11.2}$$

The assumption of a lossless medium, however, is an idealization. There is always some loss of acoustic energy when sound propagates in real media. This is the so-called *dissipation* of sound energy into thermal energy. Dissipation causes *spatial damping* of the sound waves.

The wave equation can account for small dissipation by replacing the imaginary term, the *phase coefficient*[1], $j\beta$, by a complex one. This term is the complex *propagation coefficient*

$$\underline{\gamma} = \breve{\alpha} + j\beta, \tag{11.3}$$

where $\breve{\alpha}$ is the *damping coefficient* that describes spatial damping[2]. The resulting form of the wave equation is

---

[1] As mentioned before, we prefer the letter symbol β to k ... wave number

[2] Recall that we had already introduced $\breve{\alpha}$ with the exponential wave in Section 8.3. There, however, the damping was caused by geometric expansion of the wave and not by dissipation

$$\frac{\partial^2 \underline{p}}{\partial x^2} - \underline{\gamma}^2 \underline{p} = 0,$$ (11.4)

and has the plane-wave solution

$$\underline{p}_\rightarrow (x) = \underline{p}_\rightarrow e^{-\check{\alpha} x} e^{-\mathrm{j}\beta x}.$$ (11.5)

The mathematic description mentioned above is analogous to wave propagation in loss-afflicted electric transmission lines. In fact, the relevant electric and acoustic equations are homomorphic. Figure 11.1 illustrates the equivalent electric circuit with the resistance load, R', the inductance load, L', the susceptance load, G', and the capacitance load, C'.

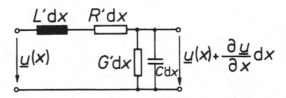

Fig. 11.1. Equivalent circuit for spatially-damped plane-wave propagation

The wave equation and the complex propagation coefficients for the equivalent electric system are known from electrical transmission theory as

$$\frac{\partial^2 \underline{u}}{\partial x^2} - \underline{\gamma}^2 \underline{u} = 0 \quad \text{and, consequently,}$$ (11.6)

$$\underline{\gamma} = \sqrt{(R' + \mathrm{j}\omega\, L')(G' + \mathrm{j}\omega\, C')}.$$ (11.7)

This leads to the characteristic impedance of the transmission line, the so-called line impedance, $\underline{Z}_\mathrm{L}$, namely,

$$\underline{Z}_\mathrm{L} = \sqrt{\frac{R' + \mathrm{j}\omega\, L'}{G' + \mathrm{j}\omega\, C'}}.$$ (11.8)

Please note that, in general, the line impedance $\underline{Z}_\mathrm{L}$ becomes complex when losses are involved. This holds for acoustic wave propagation as well – as will be shown in Section 11.2.

In acoustics, the term *dissipation* does not have the same meaning as *absorption*. Absorption means that sound disappears from a specified space through a boundary. The sound energy that leaves the space is called absorbed, no matter whether it dissipates or just transmits into another space. The current chapter will deal with absorption in Section 11.4. Absorption is a phenomenon of high technical relevance – for example, in room-acoustics practice.

11.1 Dissipation During Sound Propagation in Air

The assumptions for lossless sound propagation, as put forward in Section 7.1 are as follows,

- No inner friction, which means no viscosity
- Negligible thermal conduction
- Behavior as a perfect gas

It goes without saying that these assumptions are not strictly valid in real fluids. Therefore we have to consider the following three reason for dissipation.

- *Viscosity* ... Real media, such as air, show at least some viscosity. As a result, the periodical compressions and expansions of the medium go along with losses due to friction

- *Thermal conduction* ... There is transfer of thermal energy from the compressed and thus warmer zones to the expanded and thus cooler ones. In consequence, the compression is no longer strictly adiabatic and, hence, pressure differences are attenuated. This leads to dissipation

- *Deviation from perfect gas* ... molecular dissipation. Most gases are composed from multi-atomic molecules. For example, air contains O_2, N_2 and H_2O. These molecules do not only have a translatory degrees of motional freedom but also such as vibratory and rotatory degrees of freedom. Energy from translatory to, for example, vibratory and/or rotatory motion occurs with some delays, and effect which is called *relaxation*. This causes a deviation from adiabatic compression since energy is taken from translatory motion and returned to it at a different instance in time

It is known that all three kinds of dissipation in free sound propagation in air are proportional to the second power of the frequency, f^2. The molecular dissipation is paramount in the audio frequency range of 16 Hz–16 kHz – mainly due to vibratory modes of the molecules.

In Table 11.1 we present some estimates for the damping coefficient, $\breve{\alpha}$, in air. According data in the literature show profound variances.

Table 11.1. Damping coefficients of air in two meteorological conditions (values given in Neper – refer to Section 1.6)

Frequency, f	0.5	1	2	4	8	16	kHz
$\breve{\alpha}$ (10° C, 70 %)	0.23	0.42	1.04	3.40	12.30	39.96	10^{-3} Np/m
$\breve{\alpha}$ (20° C, 50 %)	0.32	0.57	1.13	3.20	11.00	38.60	10^{-3} Np/m

Dissipation During Sound Propagation in Tubes

In contrast to free propagation, viscosity and thermal transfer can gain in relevance when the sound propagates along a duct. In a tube with rigid walls made of heat-conducting material, one has to consider the following two effects.

- *Viscosity* ... The walls of the tubes are rigid compared to the air. Hence the particle velocity, $\underline{v}$, is zero directly on the surface of the wall. Yet, at a little distance from the wall the velocity already acts as in the progressive wave. Consequently, we have a strong velocity gradient in the radial direction within a thin boundary layer. Although the viscosity of air is low, the frictional losses occurring in this way become considerable

- *Thermal conduction* ... With the walls conducting heat well, dissipation due to heat transfer is likely. These effects are certainly larger than in a free wave as there the compressed and expanded sections are separated by quarter-wavelength air distances, namely, $\lambda/4$. But now, due to well-conducting walls, these sections are in a more direct contact

Both effects are proportional to (a) the ratio of circumference, U, and cross-sectional area, A, and (b) the root of the frequency, $\sqrt{\omega}$. The following approximation has been derived by *Cremer* for so-called *wide tubes*, that is, tubes where the boundary layer is small compared to the diameter, namely,

$$\breve{\alpha} \approx 6.7 \cdot 10^{-6} \, (U/A) \sqrt{\omega}/\text{Hz} \quad \dots \text{ for wide tubes}. \tag{11.9}$$

It has been shown that viscosity accounts for $\approx 2/3^{\text{rd}}$ of the damping coefficient resulting from this formula[3].

11.2 Sound Propagation in Porous Media

From the previous discussion it is easy to comprehend that sound propagation in media that are perforated by many narrow tubes and connected cavities, must be profoundly damped. Media of this consistency are called porous[4]. With the set as shown in Fig. 11.2, the characteristic flow resistance, Ξ, of porous media can be measured, though, to be precise, only for the "static"(steady) case of a continuous gas flow. Ξ is defined as follow,

$$\Xi \, v_a = -\frac{\Delta p}{\Delta x} \approx -\frac{\partial p}{\partial x}\Big|_{\text{x}}. \tag{11.10}$$

---

[3] In the analogous circuit of Fig. 11.1, viscosity corresponds to R' and thermal conduction to G'

[4] Note that a medium with fully enclosed, that is, mutually unconnected cavities is *not* porous. Artificial foams can be produced both ways, that is, either with or without connections between the internal voids (gas bubbles)

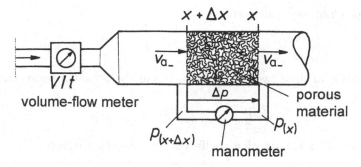

Fig. 11.2. Set up for measuring the static flow resistance, $\varXi$

In a strongly simplified model introduced by *Rayleigh*, the porous medium is replicated as a rigid skeleton that is perforated by perpendicular narrow ducts – illustrated in Fig. 11.3. The ratio of the duct area inside the probe, A_i, to the total area outside the probe, A_a, is called *porosity*, σ, namely,

$$\sigma = \frac{A_i}{A_a} < 1 . \tag{11.11}$$

The volume velocities inside, $\underline{q}_i$, and outside the probe, $\underline{q}_a$, are the same due to mass conservation, that is $\underline{q}_i = \underline{q}_a$. This means that the respective particle velocities, v_i and v_a, relate as follows.

$$\sigma\, v_i = v_a \quad \text{and} \quad \sigma\, \varXi\, v_i = -\frac{\partial p}{\partial x} . \tag{11.12}$$

Euler's equation is now complemented by this loss term while the continuity equation stays the same. We thus get the following pair of linear differential wave equations for inside porous media,

$$-\frac{\partial p}{\partial x} = \varrho\, \frac{\partial v_i}{\partial t} + \sigma\, \varXi\, v_i \quad \text{and} \quad -\frac{\partial v_i}{\partial x} = \kappa\, \frac{\partial p}{\partial t} . \tag{11.13}$$

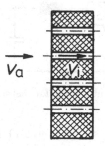

Fig. 11.3. *Rayleigh*'s model of porous materials

In complex notation this reads

$$-\frac{\partial p}{\partial x} = j\omega\,\varrho\,\underline{v}_i + \sigma\,\Xi\,\underline{v}_i \quad \text{and} \quad -\frac{\partial \underline{v}_i}{\partial x} = j\omega\,\kappa\,\underline{p}. \tag{11.14}$$

Combination of these two equations results in the 2nd-order wave equation for porous media,

$$\frac{\partial^2 p}{\partial x^2} - (j\omega\varrho + \sigma\,\Xi)(j\omega\,\kappa)\,\underline{p} = 0. \tag{11.15}$$

For the complex propagation coefficient we subsequently get

$$\underline{\gamma} = \sqrt{(j\omega\varrho + \sigma\,\Xi)(j\omega\,\kappa)}, \tag{11.16}$$

and for the characteristic impedance,

$$\underline{Z}_w = \frac{\underline{p}_{i\rightarrow}}{\underline{v}_{i\rightarrow}} = \sqrt{\frac{j\omega\,\varrho + \sigma\,\Xi}{j\omega\,\kappa}}. \tag{11.17}$$

For *low frequencies* the complex propagation coefficient is approximately proportional to the root of the frequency, that is

$$\underline{\gamma} \approx \sqrt{\sigma\,\Xi\,\omega\,\kappa}\;e^{j45°} \quad \text{and, thus,} \quad \breve{\alpha} = \sqrt{\sigma\,\Xi\,\omega\,\kappa} \sim \sqrt{f}. \tag{11.18}$$

For *high frequencies* the propagation coefficient asymptotically reaches a limiting value that is real, as can be shown by eliminating $\breve{\alpha}$ from the formula for $\underline{\gamma}$, namely,

$$\breve{\alpha}_{asymp} = \frac{\sigma\,\Xi}{2\,\varrho\,c}. \tag{11.19}$$

To be sure, the *Rayleigh* model mimics porous media only very roughly. For practical applications the porosity σ can be determined by measurements. Furthermore, it has to be considered that the air in cavities inside the medium increases the volume compliance, κ, but is not accelerated. This is accounted for by a structural factor, $\chi \leq 1$, which can be determined by dynamic measurements. This factor is introduced into the wave equation by substituting ϱ by $\varrho\,\chi$.

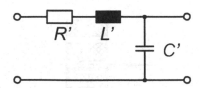

Fig. 11.4. Equivalent circuit for inside a porous medium

The equivalent electric circuit – shown in Fig. 11.4 – is appropriate for the wave equation inside porous media as derived above. Obviously, only viscosity and no thermal conduction has been included in the modeling.

11.3 Reflection and Refraction

Reflection and *refraction* are important phenomena in wave propagation. They can be treated together by assuming a situation where a wave hits the boundary between two media, medium 1 and medium 2, with different speeds of sound, c_1 and c_2. If we take both the boundary and the wave as infinite in space, the situation can be depicted in one plane – shown in Fig. 11.5.

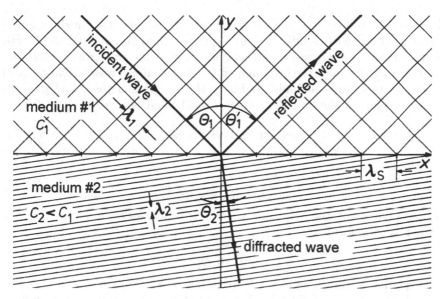

Fig. 11.5. Wave propagation at the boundary between two fluid media

Reflection

The incoming wave can be split into two orthogonal components, one of which propagates in $-y$-direction and the other one in $+x$-direction. The x-component is not affected by the boundary in any way. The $-y$-component, however, inverts its direction at the boundary from $-y$ to $+y$, due to reflection, while maintaining its speed c_1.

The components of the phase coefficient in x-direction, β_x, are identical for the incoming and reflected waves, in other words

$$\beta_1 \sin \Theta_1 = \beta_1 \sin \Theta_1' \,, \tag{11.20}$$

or, with $\beta_1 = 2\pi/\lambda_1$,

$$\frac{\lambda_1}{\sin \Theta_1} = \frac{\lambda_1}{\sin \Theta_1'} = \lambda_T \,, \tag{11.21}$$

where λ_T is the wavelength of the trace of the wave along the boundary, the so-called *trace wavelength*.

Since, due to the boundary condition $\underline{p}_1(x) = \underline{p}_2(x)$, the trace wavelength is identical for both the incident and reflected waves, we directly arrive at the law of reflection that states that the incoming angle is identical to the outgoing one, or

$$\Theta_1 = \Theta_1' . \tag{11.22}$$

Refraction

The non-reflected part of the incoming wave transmits into medium 2 and propagates there with a different sound speed, c_2, and at a different angle, Θ_2. At the boundary, the phase coefficient and, hence, the trace wavelengths are again identical for both the incoming and the transmitted waves. The physical reasons for this are that fluids exhibit no shear, and that the sound pressure is homogeneous at the boundary. Consequently, one can write

$$\beta_1 \sin \Theta_1 = \beta_2 \sin \Theta_2 , \tag{11.23}$$

and, thus,

$$\frac{\lambda_1}{\sin \Theta_1} = \frac{\lambda_2}{\sin \Theta_2} = \lambda_T . \tag{11.24}$$

By multiplying with the frequency, which is identical for both waves, we arrive at the following form,

$$\frac{c_1}{c_2} = \frac{\sin \Theta_1}{\sin \Theta_2} . \tag{11.25}$$

This equation is known as the refraction law of *Snellius* (or *Snell*).

If the conditions of *Snellius'* law are not met, no sound is transmitted into medium 2. This may, for example, happen for shallowly oblique (grazing) incidence of sound from air to water. In such a case, which is called *total reflection*, there is no refracted wave at all.

11.4 Wall Impedance and Degree of Absorption

This section deals with the wave effects at the boundary between two fluid media in more detail and, by appropriate extrapolation, will also approximately cover the effect of waves encountering a wall.

Boundary Between Two Fluids

As a starting point we take a case as depicted in Fig. 11.5. The characteristic impedances of the two media are $\underline{Z}_{w_1} = \varrho_1 c_1$ and $\underline{Z}_{w_2} = \varrho_2 c_2$. We now consider the boundary to medium 2 as a wall with the wall impedance $\underline{Z}_{wall} = \underline{Z}_{w_2}$.

Following Section 7.5, the reflectance $\underline{r}$ for perpendicular incidence from medium 1 on medium 2 is found to be

$$\underline{r}_\perp = \frac{\underline{Z}_{\text{wall}} - \underline{Z}_{w_1}}{\underline{Z}_{\text{wall}} + \underline{Z}_{w_1}} = \frac{\varrho_2\, c_2 - \varrho_1\, c_1}{\varrho_2\, c_2 + \varrho_1\, c_1}. \tag{11.26}$$

For oblique incidence, only those components of the particle velocity that are perpendicular to the boundary are reflected or transmitted. In acoustics, the wall impedance, $\underline{Z}_{\text{wall}}$, is defined as the ratio of the total sound pressure and the normal velocity in the wall. Accordingly, we get

$$\underline{Z}_{\text{wall}}(\varTheta_1) = \frac{\underline{p}_{2\rightarrow}}{\underline{v}_{2\rightarrow} \cos \varTheta_2} = \frac{\underline{Z}_{\text{wall}}(\varTheta = 0)}{\cos \varTheta_2}, \tag{11.27}$$

where application of *Snellius'* law yields

$$\cos \varTheta_2 = \sqrt{1 - \left(\frac{c_2}{c_1}\right)^2 \sin^2 \varTheta_1}. \tag{11.28}$$

In conclusion, the reflectance for oblique incidence results in

$$\underline{r}(\varTheta_1) = \frac{\underline{Z}_{\text{wall}}(\varTheta_1) - \frac{\varrho_1\, c_1}{\cos \varTheta_1}}{\underline{Z}_{\text{wall}}(\varTheta_1) + \frac{\varrho_1\, c_1}{\cos \varTheta_1}} = \frac{\underline{Z}_{\text{wall}}(\varTheta_1) \cos \varTheta_1 - \varrho_1\, c_1}{\underline{Z}_{\text{wall}}(\varTheta_1) \cos \varTheta_1 + \varrho_1\, c_1}. \tag{11.29}$$

This simple case shows already that the wall impedance $\underline{Z}_{\text{wall}}$, in general, is a function of the incoming oblique angle $\varTheta_1$. Further, in (11.29) it is weighted with $\cos \varTheta_1$.

Locally Reacting Boundaries and Walls

With $c_2 \ll c_1$, the angle of the refracted sound, $\varTheta_2$, approaches zero. Ergo, the refracted wave propagates perpendicularly away from the boundary. The wall impedance, then, no longer depends on the angle of incidence of the incoming sound. This means that $\underline{Z}_{\text{wall}} \neq f(\varTheta_1)$.

The same holds for wall structures where adjacent wall elements are not coupled to each other because this causes that all movements parallel to the wall are suppressed. Walls which react in this way are said to be *locally reacting*. The *Rayleigh* model of porous materials shows this feature, but many technologically relevant porous materials and wall constructions do actually behave approximately the same.

Deriving the Degree of Absorption from Wall Impedances

The acoustic power that is absorbed at a boundary is described by the *degree of absorption*, α, that has been introduced in Section 7.5 as

$$\alpha = 1 - |\underline{r}|^2 . \tag{11.30}$$

If we plot the wall impedance, $\underline{Z}_{\text{wall}}$, in the complex $\underline{Z}$-plane – as depicted in Fig. 11.6 – we find the trajectories for constant $|\underline{r}|$ and constant α to be so-called *Appollonian* circles[5].

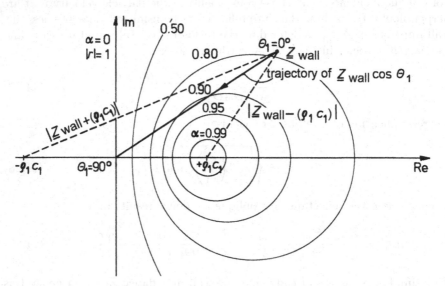

Fig. 11.6. Trajectories of constant degrees of absorption, α, in the complex Z-plane

An *Appollonian* circle is the geometric location of all positions from which the ratio of the distances to two reference points is constant. Conveniently, this is exactly what is required by the formula

$$|r(\Theta_1)| = \left| \frac{\underline{Z}(\Theta_1) - \varrho_1\, c_1}{\underline{Z}(\Theta_1) + \varrho_1\, c_1} \right| . \tag{11.31}$$

The trajectories of $\underline{Z}(\Theta_1) = \underline{Z}_{\text{wall}}\, \cos\Theta_1$, for $0 \leq \Theta \leq 90°$, are straight lines that can be used to determine the degree of absorption, α. In Fig. 11.7, two choices are shown as examples. To understand how they have been derived, think of the α-circles as isohypses – that is, lines of equal height.

It becomes clear that α approaches zero for wall-parallel incidence. For $|\underline{Z}_{\text{wall}}| > \varrho_1\, c_1$, there exists an optimum match in the sense that α assumes a maximum at a specific Θ_1. For $|\underline{Z}_{\text{wall}}| < \varrho_1\, c_1$, the best possible match and, thus, the highest degree of absorption are achieved for perpendicular sound incidence.

---

[5] By conformal transformation of the complex $\underline{Z}$-plane into the complex $\underline{r}$-plane we get a so-called *Smith* chart, which could also be used in this context

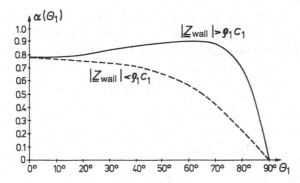

Fig. 11.7. Absorption as a function of angle Θ_1

In practical room acoustics what is mostly of interest is the degree of absorption averaged over all possible angles of incidence – so-called *diffuse-field incidence* or *random incidence*. As will be derived in Section 12.5, this diffuse-field incidence is given by the integral

$$\overline{\alpha} = \int_0^{\pi/2} \alpha(\Theta_1) \sin(2\,\Theta_1)\,d\Theta_1 . \tag{11.32}$$

Note that for locally reacting walls this quantity will never reach one.

11.5 Porous Absorbers

Porous absorbers are very important for practical applications. They are, for example, built from fibrous materials such as fabric, mineral wool or cocos fibre that is compressed into mats or plates as well as from porously extruded artificial foams. To arrive at an estimate of the absorptive behavior, the *Rayleigh* model – see Section 11.4 – is useful again. When considering perpendicular sound incidence and substituting $\underline{v}_i$ by $\underline{v}_a$ in (11.17) $\underline{Z}_{\text{wall}}$, the wall impedance is found to be

$$\underline{Z}_{\text{wall}} = \frac{1}{\sigma} \sqrt{\frac{j\omega\varrho + \Xi\,\sigma}{j\omega\,\kappa}} . \tag{11.33}$$

For high and low frequencies the following approximations hold. For *high frequencies* we get

$$\underline{Z}_{\text{wall}} \approx \frac{1}{\sigma} \sqrt{\frac{\varrho}{\kappa}} = \frac{1}{\sigma}\,\varrho\,c , \tag{11.34}$$

which is real and does not depend on frequency. For *low frequencies*, however, we find a complex and frequency-dependent relationship, namely,

$$\underline{Z}_{\text{wall}} \approx \frac{1}{\sigma} \sqrt{\frac{\Xi\,\sigma}{\omega\,\kappa}}\, e^{-j\,45°} . \tag{11.35}$$

The trajectory of $\underline{Z}_{\text{wall}}$ in the $\underline{Z}$-plane is shown in Fig. 11.8. To discuss the course of α, it is helpful to think of the α-circles as isohypses again. Accordingly, for $\alpha(\omega)$ one gets a monotonically increasing curve with a maximum of

$$\alpha_{\max} = \frac{4\sigma}{(1+\sigma)^2} \, . \tag{11.36}$$

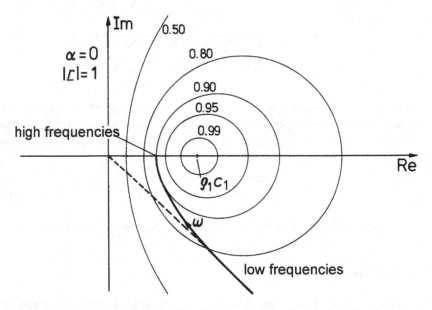

Fig. 11.8. Trajectory of $\underline{Z}_{\text{wall}}$ in the $\underline{Z}$-plane for porous absorbers

These conditions are valid for an infinitely thick layer of porous material. If the absorber thickness is finite and the material is placed upon a rigid wall, part of the energy will be reflected and re-transmitted – after having passed the absorbing material a second time.

The situation can be illustrated by regarding that in front of the wall a standing wave will develop. Directly upon the wall the perpendicular component of the particle velocity is zero, that is $\underline{v}_\perp = 0$. Consequently, the absorber is ineffective at this point – see Fig. 11.9. Thus, for low frequencies, a finite layer of material will provide less absorption than an infinitely thick one. The material is actually best exploited when positioned at a distance to the reflecting, rigid wall. α will arrive at a relative maximum whenever the absorptive layer is in a velocity maximum. This arrangement is frequently used in praxi. Figure 11.10 schematically shows the three cases as discussed above.

Real porous absorbers are not well-represented by the *Rayleigh* model, even with the structural factor, χ, being employed. Among other things, it is

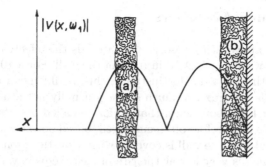

Fig. 11.9. Illustrating absorption by a layer of porous material in front of a hard wall, **(a)** with air gap, **(b)** directly on the wall

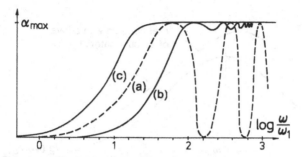

Fig. 11.10. Absorption of porous material as a function of frequency – schematic **(a)** finite layer with air gap, **(b)** finite layer directly on wall, **(c)** infinite thickness

often not clear whether the absorber arrangement is actually locally reacting. This can be enforced by *cassetting* – sketched in Fig. 11.11 – although for compressed mineral wool this is usually not necessary.

To achieve a very high α, one can try to enhance the effective absorptive area, for instance, with porous wedges. Audience, by the way, is also a considerable absorber.

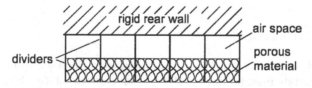

Fig. 11.11. Cassetted porous absorber

11.6 Resonance Absorbers

As we have seen above, with absorptive materials the effective layer must be placed a quarter wavelength, $\lambda/4$, in front of the wall. For a 100-Hz frequency with a wavelength of 3.40 m, for instance, this would mean a placement of 85 cm before walls. In praxi, so much space is usually not available.

Especially for low frequencies, that is, for so-called *bass traps*, a different absorber principle is therefore often employed, based on *resonance absorption*. To this end, the absorptive wall is covered with acoustic resonators, the input impedance of which is very low at their resonance frequency. Such resonators can be efficiently positioned within enclosed spaces, for example, in the corners of a room.

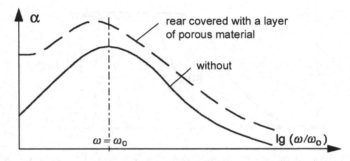

Fig. 11.12. Absorption of resonance absorbers as a function of frequency

The principle frequency relationship of α for resonance absorbers is plotted in Fig. 11.12, both with and without additional porous material.

Technical data for diffuse sound incidence for practical application are usually taken from the literature or directly from the suppliers. In the following, we only present fundamental theoretic ideas. Two types of resonance absorbers are in use, *Helmholtz absorbers* and *membrane absorbers*, which also exist in combined and/or integrated form.

Helmholtz Absorbers

Plates with holes or slits in them are placed at a distance from a wall. Absorptive materials may be put on the rear side of the plates. Figure 11.13 (a) illustrates the arrangement. For the wall impedance we get

$$\underline{Z}_{\text{wall}} = \frac{\underline{p}_{\rightarrow}}{\underline{v}_{\rightarrow}} = \underline{Z}_{\text{a}}\, A = j\omega\, m'' + \frac{1}{j\omega\, n''} + r'', \qquad (11.37)$$

where $r'' = \varXi\, b_1$ is the area-specific resistance, $n'' = \kappa\, b_2$ the area-specific compliance, and $m'' = \varrho\, (b_3 + \eta)\, \sigma$ the area-specific mass. Thereby $\sigma = \pi\, r^2 / b_4^2$

is the degree of perforation (porosity). η is a correction factor (mouth correction) which considers that at the mouth of a hole or slit more air mass is moved than is actually inside the mouth. An estimate for circular holes is $\eta = 1.6\,r$.

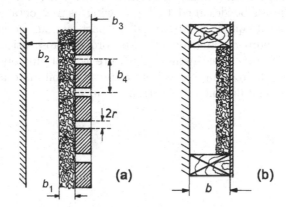

Fig. 11.13. Resonance absorbers. (a) *Helmholtz* absorber, (b) membrane absorber

Membrane Absorbers

These absorbers are built with co-vibrating membranes in front of an air gap before a wall. The membranes can, for example, be plates or other mass-afflicted materials, such as foils, and can additionally be loaded with weight to decrease the resonance frequency. Absorptive material may also be put into the air space. The arrangement is depicted in Fig. 11.13 (b). Numerous different built forms have been applied, including such with more than one membrane layer.

Although bending waves of the membranes are certainly possible, for rough calculations one usually assumes wall-perpendicular movements only. The reason for this assumption is that reacting forces due to bending of the material are usually negligible compared to those due to the stiffness of the air cushion. m'' is the area-specific mass of the plate, and $n'' = \kappa\, b$ is its compliance. The area specific resistance, r'', is hard to estimate. It contains losses within the plate.

Microperforated Absorbers

A special kind of membrane absorbers uses *microperforated* membranes. These absorbers are built with perforated thin panels or foils in front of an air gap before a rigid wall, similar to what is shown in Fig.11.13 (a), but without absorptive materials on the rear side of the panel. The perforations in the thin panel or foil is in the sub-millimeter range (diameter 0.1–1 mm) so as to provide high acoustic resistance but low area-specific acoustic-mass reactance. This is necessary for wide-band absorber. Besides the microperforated panel or foil, there is no additional fibrous, porous materials.

Microperforated absorbers are of resonant type. The bandwidth of single-panel absorbers can be designed to be as wide as 1–2 octaves. With two different resonant frequencies about 20 % apart as can be realized with double-layered microperforated panels, even broader absorption bandwidths are achieved. Yet, the most intriguing feature is that microperforated absorbers can be made from a great variety of panel or foil materials, including thin metal sheets and flexible and/or translucent foils.

12

Geometric Acoustics and Diffuse Sound Fields

So far in this book we dealt with sound propagation in terms of the wave equation. This procedure becomes very complicated, however, when treating sound fields inside rooms with complicated shapes like concert halls or churches. An approximate method called *geometrical acoustics* is often useful in these cases.

This method considers sound propagation in terms of so-called *sound rays*. Sound rays were already introduced in Section 10.4, where a ray symbol was used to designate the wave bundle that emerges from a circular hole in a rigid wall when $R \gg \lambda$. The idea is that the wave bundle propagates along a straight line like a ray of light.

The concept of rays is mathematically achieved by maintaining plane areas of constant phase and letting the wavelength go to zero. In praxi, wave propagation can be approximated by rays when the following condition is met. The wavelength of the sound under consideration must be small compared to the linear dimensions of boundary areas and obstacles. Diffraction is neglected in this view.

The energy density, W'', within a ray is equal to the energy density in a plane propagating wave. To compute its amount, we consider a wave bundle propagating through an area of $1/m^2$ for $1\,m$ – see Fig. 12.1 for illustration. The energy density, then, is the active power, $\overline{P}$, times the traveling time, $t_1 = (1/c) \cdot 1\,m$, divided by the volume of $1\,m^3$ – or, in mathematical terms,

$$W''_{\text{ray}} = \underbrace{\underbrace{|\overrightarrow{I}| \cdot 1\,m^2}_{\overline{P}} \underbrace{\frac{1}{c} \cdot 1\,m}_{t_1} \underbrace{\frac{1}{1\,m^3}}_{V^{-1}}}_{W} = \frac{|\overrightarrow{I}|}{c} = \frac{\overline{I}}{c}. \tag{12.1}$$

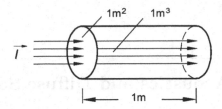

Fig. 12.1. Sound rays propagating through a unit volume

Rays are usually considered to be incoherent so that their energy densities superimpose when they meet[1]. The sum up of the rays is

$$\sum W''_{ray} = \frac{\sum |\vec{I}|}{c} = \frac{\bar{I}_\Sigma}{c}. \tag{12.2}$$

The assumption that the rays are incoherent is valid for most broadband signals like speech or music, assuming that the rays have traveled different distances from the source. This is not the case, however, for impinging and reflected waves close to reflecting surfaces. Incoherence can also not be assumed for narrow-band or pure-tone signals.

12.1 Mirror Sound Sources and Ray Tracing

The behavior of rays at plane reflecting surfaces is particularly relevant for geometrical acoustics. Plane means here, that any unevenness of the surface is small compared to the wavelengths of the sound considered. The reflection law, $\Theta_1 = \Theta'_1$, holds, and may even be applied to slightly curved planes as long as the curvature is small compared to the wavelength – shown in Fig. 12.2.

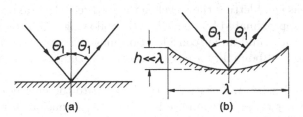

Fig. 12.2. Reflection of sound rays at **(a)** planes, and **(b)** moderately curved plates

It is particularly useful to treat sound propagation by using rays when studying room acoustics and outdoor sound propagation over longer distances – as often necessary in connection with noise control problems. When using the

---

[1] Consult Section 1.6 for incoherent superposition

concept of sound rays, relevant rules and laws from optics can directly be applied. The length of a ray is proportional to its traveling time, making it possible to not only determine the direction of sound propagation, but also the arrival times of different rays at a certain point of interest.

Reflection on plane surfaces can be depicted by *mirror sources* (*virtual sources*) – shown in Fig. 12.3. The mirror source, q_m, and the primary source, q_0, simultaneously send out identical sound fields. The combination of these sound fields on the surface produces a reflected wave that fulfills the boundary condition for full reflection, namely, the normal component of the particle velocity, $v_\perp$, being zero.

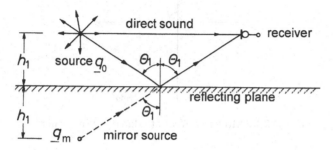

Fig. 12.3. Mirror sound sources emanating from reflection at a plane

In Figure 12.4, it is assumed that both sources transmit a short sound impulse at the initial time, t_0. The figure shows the wave fronts of both the primary and the reflected sounds and illustrates how the second wave front arrives at the receiver later than the first.

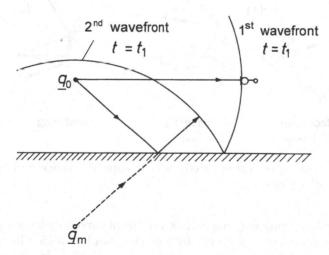

Fig. 12.4. Wave fronts of both the primary and the reflected sounds

Investigations into the relative arrival times of reflections are important, especially since reflections that arrive at the receiver with a delay may cause the perception of disturbing echoes, which should be avoided in room acoustics.

The actual perceptual *echo threshold* is dependent on the character of the sound. It is about 50 ms for running speech, larger for music and shorter for impulses.

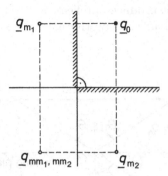

Fig. 12.5. Mirror sources at edges and in corners

Complications may arise for higher order reflections on the edges and in the corners of rectangular spaces. Mirror sources may coincide spatially. This is illustrated in Fig. 12.5 where the 2nd-order mirror sources, $\underline{q}_{\mathrm{mm}_1}$ and $\underline{q}_{\mathrm{mm}_2}$, coincide in the rectangular corner.

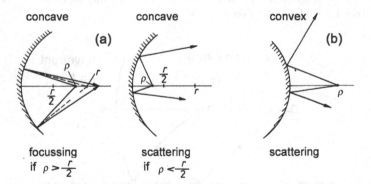

Fig. 12.6. Focussing or scattering effects (ρ denotes the source position) **(a)** at concave, **(b)** at convex surfaces

Focussing or scattering may happen as a result of curved surfaces, – shown in Fig. 12.6. Unwanted echoes caused by focussing can be avoided by modifying the form of the reflecting surface, employing irregular reflecting structures with linear dimensions that are on the order of the wavelength, or by covering the surfaces with sound-absorbing materials.

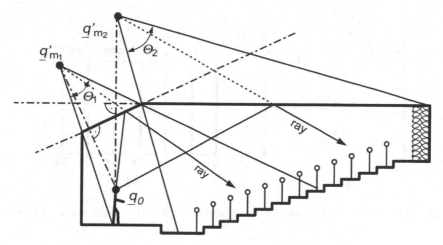

Fig. 12.7. Guiding sound via the ceiling of an auditorium

Figure 12.7 presents an application example for geometrical acoustics. In an auditorium, the sound from a speaker, $\underline{q}_0$, is guided to the audience via ceiling reflections. $\underline{q}_{m_1}$ and $\underline{q}_{m_2}$ are the mirror sound sources representing the tilted and horizontal parts of the ceiling, respectively. Two sample rays are depicted for illustration. The two mirror sources illuminate the spatial sections Θ_1 and Θ_2. The rear wall is made absorptive to avoid audible echoes.

As to the construction of the graph please note that the mirror sources are positioned perpendicularly to the reflecting surfaces at the same distance to the surface as the original source, yet, outside the room under consideration. The rays originating from them are restricted to the spatial sector defined by the individual reflecting surfaces concerned.

12.2 Flutter Echoes

We will now consider a case involving a highly directional sound source, $\underline{p}_1$, between two parallel walls a distance, l, apart. The source emits a short sound-pressure impulse directed perpendicularly toward one of the walls and propagating like a ray – shown in Fig. 12.8 (a). The two walls may be slightly absorbent, characterized by a degree of absorption, α. A microphone close to the position of the source would records a signal as schematically plotted in Fig. 12.8 (b).

If the interval between the individual impulses at the receiver, $\tau = l/c$, is larger than the echo threshold, the impulses become perceptible as a series of individual echoes, called *flutter echo*. Flutter echoes should be avoided in room acoustics. This can be accomplished by slightly tilting the two walls by $> 5°$ or by making their surfaces absorbing or scattering.

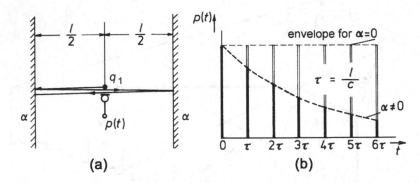

Fig. 12.8. Multiple reflections between parallel walls – the origin of flutter echoes

The envelope of the impulse series decreases exponentially for $\alpha > 0$, meaning that the ray looses a given percentage of its energy whenever a reflection takes place[2]. The following variables are involved in the flutter echo situation, with W_0'' being the energy density of the impinging ray,

$$W_1'' = (1 - \alpha)\, W_0'' \quad \dots \text{ energy density after the 1st reflection}$$
$$W_n'' = (1 - \alpha)^n\, W_0'' \quad \dots \text{ energy density after the } n\text{th reflection}$$

With l being the traveled distance between two reflections, we arrive at a temporal *reflection density*, n', that is the number of reflections per time, of

$$n' \approx c/l. \tag{12.3}$$

Please note that n' is correct up to a rest that is negligible when $l \ll 340\,\text{m}$ holds. 340 m is the distance that sound propagates in air in one second.

For large n, the expression for W_n'' can be substituted with a monotonic function as follows,

$$W''(t) = W_0''\, (1 - \alpha)^{n't} = W_0''\, (1 - \alpha)^{(c/l)t}. \tag{12.4}$$

The discrete energy losses are replaced in this way by continuous spatial damping, which allows the above expression to be written using $y = e^{\ln y}$, namely,

$$W''(t) = W_0'' \left[e^{\ln(1-\alpha)} \right]^{n't} = W_0''\, e^{n' \ln(1-\alpha)\, t}. \tag{12.5}$$

---

[2] The chunks of sound energy that the source sends out and that subsequently oscillate between the two walls are sometimes dubbed "sound particles"

This decreasing exponential function is actually an analytical description of the decreasing envelope in Fig. 12.8 (b).

For very small amounts of absorption, $\alpha \ll 1$, the expression can be simplified by using the serial expansion

$$-\ln(1-\alpha) = \alpha + \frac{\alpha^2}{2} + \frac{\alpha^3}{3} + \ldots \approx \alpha \qquad (12.6)$$

and truncating it after the first term, so that

$$W''(t) \approx W_0'' \, e^{-n'\alpha t}. \qquad (12.7)$$

12.3 Impulse Responses of Rectangular Rooms

We will now move beyond the case of two parallel walls and consider a rectangular (cuboid) room with six reflecting boundaries, namely, four walls, one floor and one ceiling. This room will illustrate an important rule in room acoustics that which states that the reflection density, n', increases with t^2 in many rooms.

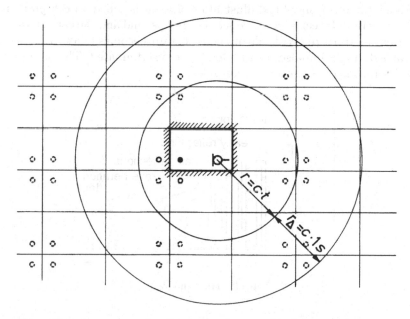

Fig. 12.9. Image sources of one sound source for a rectangular room

Figure 12.9 illustrates this concept. The figure shows the plan of a rectangular room with a single sound source in it, along with mirrored rooms of nth order with one mirror source in each of them[3].

Let V be the volume of the cuboid room. Now all of the mirror sources simultaneously transmit a sound impulse at $t = 0$. All impulses that originate from within a hollow sphere with the radius $r_\Delta = c \cdot 1\,s$, arrive at the receiver in the original room within the same interval of $1\,s$. The number of mirror sources in the hollow sphere is approximately the volume of the hollow sphere divided by the volume of the original cuboids, V, which is

$$n \approx \frac{\frac{4}{3}\pi\left[(r + r_\Delta)^3 - r^3\right]}{V}. \tag{12.8}$$

For $r \ll r_\Delta$, neglecting higher-order difference terms, r_Δ^2 and r_Δ^3, this expression approaches

$$n'(t) = \frac{4\pi\,r^2\,r_\Delta}{V \cdot 1\,s} = \frac{4\pi\,c^3\,t^2}{V}. \tag{12.9}$$

In other words, the density of the impulses arriving at the receiver is increasing with the square of expired time. The reflections also come from more directions over time, resulting in an ever more homogeneous distribution both over time and space.

Figure 12.10 is a simplified illustration of what is called an *echogram*, particulary, an impulse echogram. We see the direct sound and the early, low-order reflections as discrete event. Then the echogram becomes denser and denser, so that individual impulses can no longer be discriminated. This late part of the echogram is called *reverberation tail*.

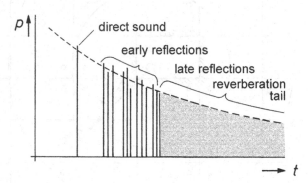

Fig. 12.10. Echogram

It is important to consider the intensity characteristics of this situation. Assuming that the sound source is a spherical source of 0th order, the active

---

[3] This is the case for rectangular rooms where many mirror sources coincide spatially due to the rectangular corners. In more irregular rooms the situation may become more complicated, particularly, when focussing occurs

intensity of the transmitted sound will drop with $1/r^2$. On the one hand, this effect can be seen for the early reflections of the echogram, but on the other hand, the reflection density increases with t^2. The active intensity measured over an interval, t_Δ, is therefore constant.

Now we have to keep in mind at this point that real instruments used to measure echograms always measure with a running time window, t_Δ, because they have low-pass characteristics. Thus, any echogram measurement would show a constant envelope as long as no absorption or dissipation occurred. Consequently, the energy in the running time window would stay constant over time. The latter is unrealistic in praxi, however, since there are always some losses. We will now show that the envelope of the reverberation tail decreases exponentially in real rooms.

12.4 Diffuse Sound Fields

From the discussion in Section 12.3, it is obvious that it is hardly possible to trace the fate of each individual sound ray, particularly in the reverberation tail. Nevertheless, it is possible to make important statements about the average fate of late reflections. Such an approach is called *statistical room acoustics*. We begin with an idealized model that adequately describes the sound field of the reverberant tail, also called the *diffuse sound field*. The model *diffuse sound field* is characterized by the following assumption, expressed in term of geometrical room acoustics.

A *diffuse sound field* is composed of many rays with the average properties of equal intensity and equal spatial distribution

This assumes that all rays, on average, have been reflected the same number of times and have, on average, traveled the same distance. It also means that the mean free-path length between two reflections is the same for all rays.

The results of statistical room acoustics are independent of room shape because only the average fate of rays is considered and described by statistical parameters.

Sound Power Impinging Upon the Walls

In a diffuse sound field composed of rays from all possible directions, the magnitude of the intensity is given by

$$\overline{I}_d = \iint_{4\pi} |\vec{I}(\Omega)|\, d\Omega = W_d'' \, c, \qquad (12.10)$$

whereby Ω is the spatial angle and $W_d'' c$ denotes the energy density. The rays that impinge on a wall from all directions transport sound power onto the wall. In the following we shall determine the total power that hits the walls perpendicularly.

We start by computing the active intensity that hits a small surface element, dA, on the wall. This intensity is obtained by integration over all differential intensities on a hemisphere of – as depicted in Fig. 12.11.

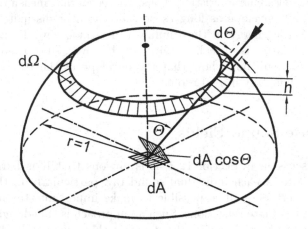

Fig. 12.11. Diffuse field impinging on a wall

The wave bundle that arrives from a spatial angle of $d\Omega$ yields the energy-density component

$$dW_d'' = \frac{W_d''}{4\pi} \, d\Omega \,, \tag{12.11}$$

where $d\Omega = 2\pi \sin\Theta \, d\Theta$ is derived from $A_{zone} = 2\pi h$, the area of a spherical zone on a unit-radius sphere. Consequently, we get

$$dI = \frac{W_d'' c}{2} \sin\Theta \, d\Theta \,. \tag{12.12}$$

The perpendicular component of the differential intensity, dI, is obtained as

$$dI_{\text{wall}} = dI \, \cos\Theta \,. \tag{12.13}$$

To get the total intensity that hits the surface element perpendicularly, we integrate over the angel Θ as follows,

$$
\begin{aligned}
dI_{\text{wall}} &= \frac{W_d'' c}{2} \int_0^{\frac{\pi}{2}} \cos\Theta \sin\Theta \, d\Theta \\
&= \frac{W_d'' c}{2} \frac{1}{2} \int_0^{\frac{\pi}{2}} \sin 2\Theta \, d\Theta = \frac{W_d'' c}{4} \,.
\end{aligned} \tag{12.14}
$$

This is apparently only $1/4^{\text{th}}$ of the total diffuse intensity. The total power impinging on a wall with total wall area, A, thus is

$$\overline{P}_{\text{wall}} = \frac{W_{\text{d}}'' \, c}{4} \, A. \tag{12.15}$$

Sound Power Absorbed by the Walls

The power absorbed by the walls is

$$\overline{P}_{\text{wall, abs}} = \frac{W_{\text{d}}'' \, c}{4} \, \overline{\alpha} \, A, \tag{12.16}$$

where $\overline{\alpha}$ is the degree of absorption for diffuse sound incidence. It can be determined by the equation

$$\overline{\alpha} = \int_0^{\frac{\pi}{2}} \alpha(\Theta) \, \sin 2\Theta \, d\Theta. \tag{12.17}$$

This expression is known as *Paris'* formula and has already been mentioned in Section 11.4. The validity of *Paris*'s formula becomes clear by realizing that the absorbed power is equal to

$$dI_{\text{wall, abs}}(\Theta) = \alpha(\Theta) \, dI_{\text{wall}}. \tag{12.18}$$

Average Reflection Density and Average Free-path Length

The energy that hits the walls during a time span of one second is

$$W_{1\text{s}} = \frac{W_{\text{d}}'' \, c}{4} \, A \cdot 1\,\text{s}. \tag{12.19}$$

The total diffuse-field energy present in a room with the volume, V, is

$$W_{\text{room}} = W_{\text{d}}'' \, V. \tag{12.20}$$

The rays transport this energy to the wall on an average of $\overline{n}'$ times per second, that is

$$\overline{n}' \, W_{\text{d}}'' \, V = \frac{W_{\text{d}}'' \, c}{4} \, A \cdot 1\,s, \quad \text{with} \quad \overline{n}' = \frac{A \, c}{4V}. \tag{12.21}$$

This leads to the expression for the mean free-path length as

$$\overline{l} = \frac{c}{\overline{n}'} = \frac{4V}{A}. \tag{12.22}$$

12.5 Reverberation-Time Formulae

When we insert the average reflection density $\bar{n}'$, and the average degree of absorption, $\bar{\alpha}$, into (12.5), which describes the decay of a multiply reflected wave bundle – see Section 12.2 – we obtain the following expression for the time-dependency of the energy density in the diffuse sound field,

$$W_{\mathrm{d}}''(t) = \sum W_{\mathrm{ray}}''(t) = W_{\mathrm{d}}''(t=0)\, e^{\left[\frac{A\,c}{4\,V}\,\ln(1-\bar{\alpha})\,t\right]}. \qquad (12.23)$$

According to *Sabine*, the time span required for the energy density to decrease to 1 millionth of its initial value, that is, by 10^6 or 60 dB, is called the *reverberation time*, T. Inserting this reverberation time into (12.23) results in

$$10^{-6}\, W_{\mathrm{d}}''(t=0) = W_{\mathrm{d}}''(t=0)\, e^{\left[\frac{A\,c}{4\,V}\,\ln(1-\bar{\alpha})\,T\right]}. \qquad (12.24)$$

In air under normal condition, that is 20° C temperature and 1000 hPa static pressure, the sound speed, c, is about 340 m/s. Using this value, we obtain

$$T = 0.163 \left[\frac{\mathrm{s}}{\mathrm{m}}\right] \frac{V}{-\ln(1-\bar{\alpha})\,A}, \qquad (12.25)$$

an expression that is known as *Eyring*'s reverberation formula. For small amounts of absorption, $\bar{\alpha} \ll 1$, this formula simplifies to

$$T \approx 0.163 \left[\frac{\mathrm{s}}{\mathrm{m}}\right] \frac{V}{\bar{\alpha}\,A}, \qquad (12.26)$$

which is preferred in praxi and known as *Sabine*'s reverberation formula.

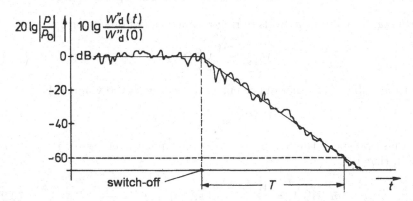

Fig. 12.12. Time trace of a reverberant noise sound after being switched off

The reverberation time, T, can be measured in many ways, including the following *switch-off method*. A room is excited with a noise source until a stationary average sound-pressure level is reached. Then the sound source is switched off, and the sound-pressure level is recorded as a function of time, producing a curve like the one shown in Fig. 12.12. The time interval between switching-off the source and the instance where the curve has decreased by 60 dB, is taken to be the reverberation time, T.

Consideration of the theory an actual reverberation plots shows that the *diffuse sound field* assumption is sufficiently valid when the following conditions are met.

- Sound absorption is well distributed about the boundaries of the room
- The shape of the room is irregular and focussing elements are particularly avoided
- Total sound absorption is small or moderate. A good check is if the ratio of room volume and equivalent absorptive area – see next paragraph – is not too small, usually $> 1\,\mathrm{m}$

The Equivalent Absorptive Area

The following expression,

$$A_\alpha = -\ln(1 - \overline{\alpha})\,A \approx \overline{\alpha}\,A, \qquad (12.27)$$

is understood to be a fictive area with a unit degree of absorption, that is 100% of absorption or $\overline{\alpha} = 1$, representing the total absorption present in the room. This fictive area is called *equivalent absorptive area*.

If there are partial areas at the boundary of a room, each with an individual absorption value of $\overline{\alpha}_i$, then the equivalent absorptive area is defined to be

$$A_\alpha = \sum_i A_i\,\overline{\alpha}_i + \underbrace{8\,\overline{\alpha}\,V}_{\text{correction}}, \qquad (12.28)$$

where A_i are the individual areas with individual degrees of absorption $\overline{\alpha}_i$.

The correction component in the formula, not derived here, is only used to account for dissipation in the transmission medium of large rooms. Values for air are given in Table 11.1.

12.6 Application of Diffuse Sound Fields

We will now introduce three well known and frequently used applications of the diffuse-sound-field model.

Reverberation-time Acoustics

The reverberation time, T, estimated with either *Eyring*'s or *Sabine*'s formula, is considered to be a relevant parameter of the *acoustic quality* of spaces, which includes, among other things, their suitability for specific performances.

Table 12.1 presents preferred values of T for various performance styles. The values are drawn from literature and refer to the 500–1000 Hz range. A moderate increase toward low frequencies is considered adequate since it is said to increase the listeners' sense of envelopment and warmth.

Table 12.1. Reverberation times

Speech	Chamber music	Opera houses	Concert halls	Organ music
0.8–1.0 s	1.4–1.6 s	1.5–1.7 s	1.9–2.2 s	2.5 s and more

The following guidance holds for speech. A too-short reverberation time produces high intelligibility but also increases the effort required from the speaker. If T is too long, auditory smearing takes place and deteriorates intelligibility. 0.8–1.0 s is a reasonable compromise for speech running at normal speed, which is roughly 50 syllables/minute.

Reverberation time is undoubtedly an important parameter of acoustic quality but certainly not the only one. Proper guidance of early reflections and avoidance of echoes are at least as important.

Measurement of Spatially Averaged Absorption

Absorption measurements are taken in *reverberation chambers*, which are rooms where a diffuse sound field has been realized using highly reflective, obliquely oriented walls and planes[4].

The equivalent absorptive area, A_{α_1}, of the empty chamber must be determined beforehand, usually by a reverberation-time measurement. With a sample of the material to be measured in the chamber, the reverberation-time measurement is then repeated. For the measured reverberation-time data, then, the equivalent absorptive area of the chamber including the sample is,

$$A_{\alpha_2} = 0.163 \left[\frac{\mathrm{s}}{\mathrm{m}}\right] \frac{V}{T}. \tag{12.29}$$

Consequently, the equivalent absorptive area of the sample results as

$$A_{\alpha,\text{sample}} = A_{\text{sample}}\, \overline{\alpha}_{\text{sample}} = A_{\alpha_2} - A_{\alpha_1}. \tag{12.30}$$

---

[4] In order to measure α for perpendicular sound incidence only, a measuring tube may be applied – see Section 7.6

Measurement of the Total Power of a Sound Source

Sound power measurements are also undertaken in the diffuse sound field of a reverberation chamber. The source is brought into the chamber and operated to transmit ongoing sound.

The sound power transmitted by the source and the sound power absorbed by the equivalent absorptive area of the reverberation chamber will reach a stationary balance as follows[5],

$$\underbrace{\overline{P}_{\text{source}}}_{\text{introduced}} = \underbrace{\frac{A_\alpha\, W_{\text{d}}''\, c}{4}}_{\text{absorbed}} \quad \text{with} \quad W_{\text{d}}'' = \frac{I_{\text{d}}}{c} = \frac{p_{\text{rms}}^2}{\varrho\, c^2} . \tag{12.31}$$

The power of the sound source, consequently, is

$$\overline{P}_{\text{source}} = p_{\text{rms}}^2\, \frac{A_\alpha}{4\, \varrho\, c} . \tag{12.32}$$

The following rules are useful when performing the measurements. A sufficient number of measuring points must be well-distributed across the room but not too close to walls or corners because full incoherence of incoming and reflected sounds can not be guaranteed in such locations. This may result in measured values that are too high.

The Critical Radius

A stationary, diffuse sound field has the same average energy density, W_{d}'', everywhere in space. Close to a sound source, however, the energy density of the direct sound may be much higher. The situation is depicted in Fig. 12.13.

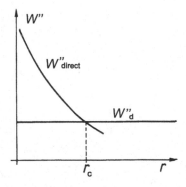

Fig. 12.13. The critical radius – equilibrium of direct- and diffuse-field energy densities with a 0th-order spherical source

---

[5] With p_{rms} being the rms-value of p – see Section 15.5 for a definition

The critical radius is the distance from a 0th-order spherical source where the direct and diffuse energy densities are just equal. Equality of the two energy densities is given at

$$W''_{\text{direct}} = \underbrace{\frac{\overline{P}_{\text{source}}}{4\pi\, r^2} \frac{1}{c}}_{I_{\text{direct}}} \overset{!}{=} W''_{\text{d}} = \underbrace{\frac{4\,\overline{P}_{\text{source}}}{A_\alpha} \frac{1}{c}}_{I_{\text{diffuse}}}, \qquad (12.33)$$

which leads to the critical radius as follows,

$$r_c = \sqrt{\frac{A_\alpha}{V}}. \qquad (12.34)$$

For directional sources the distance of equilibrium of direct and diffuse field is higher in the direction of focussed transmission. We then speak of *critical distance* or *diffuse-field distance*.

Knowing the critical radius or critical distance is useful when taking sound recordings because microphone placement within this radius or distance will predominantly render direct-sound signals. Placement outside will predominantly render diffuse-sound signals, which are auditorily perceived as *spatial impression*.

Isolation of Air- and Structure-Borne Sound

Sound isolation is the confinement of sound to a space in such a way that transmission to neighboring spaces is totally or partially prevented. Sound isolation is predominantly based on reflection caused by impedance discontinuities in possible transmission paths. Dissipation and absorption may also play a role in sound isolation, but it is usually minor. Another term for sound isolation is *sound damming* because the sound is, so-to-say, "dammed in".

Sound isolation must not be confused with *sound damping*. Damping of sound means that sound energy has been removed from a sound field by means of dissipation and/or absorption. The transmission of sound to another space is one possible method of absorption. Thus, absorption is not necessarily dissipation, the latter being transformation of acoustic/mechanic into thermal energy.

Measures of airborne and structure-born sound isolation are of particular technological relevance. Non-porous leaves or walls are typically inserted into airborne transmission paths to achieve isolation, and isolation of structure-borne sound is accomplished by inserting elastic elements (springs) or layers (resilient materials, air gaps). Sometimes heavy *interlocking masses* are also used. In every case, the goal is to create impedance discontinuities that result in reflection.

13.1 Sound in Solids – Structure-Borne Sound

An important difference between solids and fluids is that solids experience shear forces. This means that solids can store energy through both volume changes and changes of form. Solids consequently experience a number of wave types in addition to longitudinal waves since transverse movement with respect to the direction of propagation is possible. The types of waves that are actually possible in a specific case are dependent, among other things, on the specific form or the configuration of solid bodies under consideration.

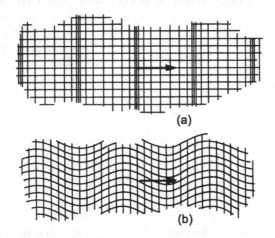

Fig. 13.1. Longitudinal, (a), and transversal waves, (b)

Infinitely extended solids only experience longitudinal density waves and transverse shear waves – shown in Fig. 13.1, but finite solids like rods or plates can also carry dilatation, surface, torsion, and bending waves – illustrated in Figs. 13.2 and 13.3.

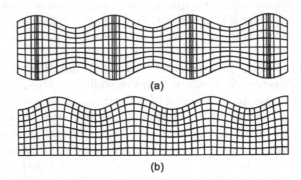

Fig. 13.2. Quasi-longitudinal (dilatational), (a), and surface waves, (b)

The different types of waves listed above couple with each other at boundaries, junctions and/or points of impact. This means that selective damping of one wave type does not prevent this kind of wave from being excited again somewhere else. For example, a rod that is perpendicularly fixed to a plate

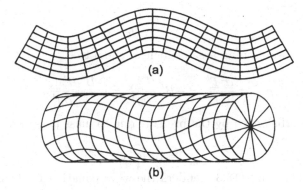

Fig. 13.3. Bending, (a), and torsion waves, (b)

and carrying an elongation wave will excite bending waves at the junction with the plate.

13.2 Radiation of Airborne Sound by Bending Waves

The specific combination of transverse and angular motion that characterizes bending waves result in considerable surface velocities and, consequently, the emission of airborne sound. This type of wave is of particular practical relevance because of this effect.

The Wave Equation for the Bending Waves

The wave equation for a lossless free bending wave in thin plates will now be derived. Energy storage in these waves is accomplished by the mass load, m'' [mass/area], and the *bending stiffness*, B'. Fig. 13.4 illustrates the situation.

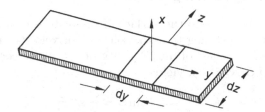

Fig. 13.4. Element of a plate with mass and bending stiffness

The bending stiffness, B', excites a torque per width, T', that is proportional to the flexion, $\partial^2 \xi_x / \partial y^2$. The following consequently holds,

$$T' = -B' \frac{\partial^2 \xi_x}{\partial y^2}.$$

(13.1)

The relative pressure on an area element is equal to

$$\underline{p}_{B'} = \frac{\partial^2 \underline{T}'}{\partial y^2} = -B' \frac{\partial^4 \underline{\xi}_x}{\partial y^4} = \frac{-B'}{j\omega} \frac{\partial^4 \underline{v}_x}{\partial y^4},$$

(13.2)

which is counterbalanced by the pressure due to mass persistence, expressed as

$$\underline{p}_{m''} = j\omega m'' \underline{v}_x.$$

(13.3)

Combining (13.2) and (13.3) renders the wave equation for bending waves, namely,

$$j\omega m'' \underline{v}_x + \frac{B'}{j\omega} \frac{\partial^4 \underline{v}_x}{\partial y^4} = 0.$$

(13.4)

The solution for propagating bending waves of this 4th-order differential equation is

$$\underline{v}_x(y) = \underline{v}_{x \to} e^{-\beta_b y}.$$

(13.5)

Inserting this solution into the wave equation result in the phase coefficient, β_b, which is equal to

$$\beta_b = \sqrt{\omega} \sqrt[4]{\frac{m''}{B'}}.$$

(13.6)

This leads directly to the phase velocity of the free bending wave, c_b,

$$c_b = \sqrt{\omega} \sqrt[4]{\frac{B'}{m''}} = \frac{\omega}{\beta_b}.$$

(13.7)

In other terms, different spectral components of the wave propagate with different speeds because the phase velocity is a function of frequency. This means that bending waves are *dispersive*.

Sound Emission by Bending Waves

The velocity perpendicular to the surface of the plate is continuous with the velocity of the plate, meaning that the solution for velocity given above also holds for the adjacent layer of air. Equality of the phase coefficients at the boundary, $\beta_b = \beta \cos \Theta = \beta_y$, yields

$$\beta_b = \frac{\omega}{c_b} = \beta \sin \Theta = \frac{\omega}{c} \sin \Theta,$$

(13.8)

from which follows

$$c = c_b \sin \Theta.$$

(13.9)

This equation is similar to the law of refraction – as introduced in Section 11.3. The equation includes the following two distinguishable cases.

- *Case I* ... Equation (13.9) is fulfilled for $c < c_b$ or, equivalently, $\beta > \beta_b$ or $\lambda < \lambda_b$. This results in radiation of airborne sound at an angle of Θ – shown in Fig. 13.5. This is because the equation

$$\beta_x = \beta \cos\Theta = \sqrt{\beta^2 - (\beta \sin\Theta)^2} = \omega \sqrt{\frac{1}{c^2} - \frac{1}{c_b^2}} = \beta_b , \qquad (13.10)$$

results in a real (non-complex) solution for β_x. This means that a component of the phase coefficient exists along the x-direction

- *Case II* ... Equation (13.9) cannot be fulfilled for $c > c_b$ because β would become imaginary and result in the real damping coefficient $\alpha_x = j(j\beta_x)$. In this case, we observe an exponential decrease in the x-direction, which physically amounts to a hydrodynamic short-circuiting near the surface. This situation is parallel to *total reflection* in refraction

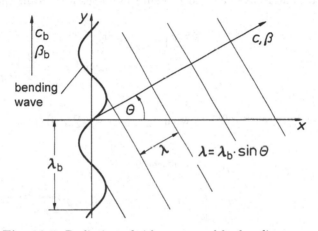

Fig. 13.5. Radiation of airborne sound by bending waves

13.3 Sound-Transmission Loss of Single-Leaf Walls

We will restrict ourselves to non-porous walls in this section. In acoustical terms *single leaf* refers to a panel in which the cross-sectional particle velocity, v_x, is identical at all points inside the leaf. This is the case in thin, solid leaves. Elongation waves inside the leaf may thus be neglected there.

Consider an infinitely extended, thin single-leaf wall in an infinitely extended fluid like an air space. The wall may have a mass load, m'', and a bending stiffness, B'.

An enforced bending wave is excited by the sound-pressure distribution in front of and behind the leaf. The excitation is controlled by the pressure difference between the two sides of the leaf shown in Fig. 13.6.

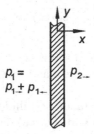

Fig. 13.6. Sound-pressure distribution at the two sides of a leaf

If we insert this differential pressure into the bending wave equation as the exciting term that represents the pressure balance, we obtain the following expression

$$[\underline{p}_{1\rightarrow}e^{-j\beta\,(\sin\Theta)y} + \underline{p}_{1\leftarrow}e^{-j\beta\,(\sin\Theta)y}] - \underline{p}_{2\rightarrow}e^{-j\beta\,(\sin\Theta)y}$$
$$= j\omega\,m''\,\underline{v}_x + \frac{B'}{j\omega}\frac{\partial^4\underline{v}_x}{\partial y^4}\,. \quad (13.11)$$

Inserting the solutions

$$\underline{v}_{x\rightarrow}(y) = \underline{v}_{x\rightarrow}e^{-j\beta\,(\sin\Theta)y} \quad \text{with} \quad \underline{v}_{x\rightarrow}(y) = \underline{v}_{2\rightarrow}(y)\cos\Theta\,, \qquad (13.12)$$

yields

$$[(\underline{p}_{1\rightarrow} + \underline{p}_{1\leftarrow}) - \underline{p}_{2\rightarrow}]\,e^{-j\beta\,(\sin\Theta)\,y}$$
$$= (j\omega m'' + \frac{B'}{j\omega}\beta^4\sin^4\Theta)e^{-j\beta\,(\sin\Theta)\,y}\,\overbrace{\underline{v}_{2\rightarrow}\,\cos\Theta}^{\underline{v}_{x\rightarrow}(y)}\,. \quad (13.13)$$

Now, since it is true that

$$\underline{v}_{2\rightarrow} = \frac{\underline{p}_{2\rightarrow}}{\varrho c} \quad \text{and} \quad \underline{v}_{2\rightarrow} = \underline{v}_{1\rightarrow} + \underline{v}_{1\leftarrow} = \frac{\underline{p}_{1\rightarrow}}{\varrho c} - \frac{\underline{p}_{1\leftarrow}}{\varrho c}\,, \qquad (13.14)$$

we may eliminate $\underline{p}_{1\leftarrow}$ and $\underline{v}_{2\rightarrow}$ and obtain

$$\frac{\underline{p}_{2\rightarrow}}{\underline{p}_{1\rightarrow}} = \frac{1}{1 + \frac{\cos\Theta}{2\varrho c}\left(j\omega\,m'' + \frac{B'}{j\omega}\beta^4\sin^4\Theta\right)}\,. \qquad (13.15)$$

Inverting and rewriting in logarithmic terms yields the so-called *transmission loss*, R,

$$R = 20\lg\left|\frac{\underline{p}_{1\rightarrow}}{\underline{p}_{2\rightarrow}}\right| \text{dB} = 20\lg\left|1 + \frac{j\cos\Theta}{2\varrho c}\left(\omega\,m'' - \frac{B'}{\omega}\beta^4\sin^4\Theta\right)\right| \text{dB}\,.$$
$$(13.16)$$

Perpendicular Sound Incidence

In the case of perpendicular sound incidence $\sin \Theta = 0$, the 2nd term in the parentheses of (13.16) vanishes. This also holds for leaves that are very compliant for bending waves, that is, for $B' \to 0$. The insertion loss, R, now approximately becomes

$$R \approx 20 \lg \left(\frac{\omega \, m''}{2 \varrho \, c} \right) \text{dB} . \tag{13.17}$$

In this case the insertion loss only depends on the mass load, m'', and increases at a rate of 6 dB per doubling of the mass load. This is called the *mass law* for walls.

Oblique Sound Incidence

An important special case related to oblique sound incidence is when the entire parenthetical term vanishes and we thus have $R = 0$. This is the case when the y-component of the phase coefficient of the sound wave in air is equal to the phase coefficient of the bending wave in the leaf, such that

$$\beta^4 \sin^4 \Theta = \omega^2 \frac{m''}{B'} \quad \text{and, thus,} \quad (\beta \sin \Theta)^4 = (\beta_b)^4 , \tag{13.18}$$

and, finally,

$$\beta \sin \Theta = \beta_b . \tag{13.19}$$

This equality of phase coefficients is called *trace matching* and can be understood as a kind of spatial resonance. A lossless wall like a window grate becomes completely transparent when these conditions occur. It follows from (13.19) that

$$\frac{\omega}{c} \sin \Theta = \sqrt{\omega} \sqrt[4]{\frac{m''}{B'}} , \tag{13.20}$$

which in turn leads to

$$\omega_c = 2\pi \, f_c = \sqrt{\frac{m'' c^4}{B' \sin^4 \Theta}} \quad \dots \text{for } \sin \Theta < 1 . \tag{13.21}$$

This critical frequency, $f_c = \omega_c/(2\pi)$, is called *coincidence frequency*. The inclusion of f_c leads to the following formula for R,

$$R = 20 \lg \left| 1 + \frac{j \cos \Theta}{2 \varrho c} \omega \, m'' \left(1 - \frac{f^2}{f_c^2} \right) \right| \text{dB} . \tag{13.22}$$

Below a *limiting coincidence frequency*, $f_{c,\text{lim}}$, trace matching becomes impossible because

$$\lambda_{\text{air}} > \lambda_b . \tag{13.23}$$

This limiting frequency is determined by $\lambda_{\text{air}} = \lambda_{\text{b}}$, which leads to

$$f_{\text{c,lim}} = \frac{1}{2\pi} \sqrt{\frac{m'' c^4}{B'}} \quad \dots \text{ for } \sin\Theta = 1, \text{ which is } \Theta = 90°. \quad (13.24)$$

The following advisements are interesting for practical purposes. To achieve high transmission loss, the coincidence frequency should be well above or below the spectral region under consideration. The coincidence frequency of leaves made of solid brick or concrete usually lies in the range of 50–100 Hz. Plywood and dry-plaster panels show coincidence frequencies of about 1–3 kHz. The limiting coincidence frequency can be increased by loading the panel with additional mass or by cutting slits into it, which makes it more compliant for bending waves. One can also try to dampen the bending waves by applying absorptive coatings or viscous internal *sandwich layers*.

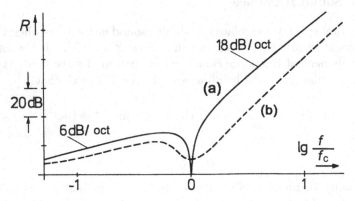

Fig. 13.7. Transmission loss as a function of frequency for a single-leaf wall. **(a)** directed oblique incidence, **(b)** random incidence

The principal frequency relationship of R as a function of f for both directional and random (diffuse-field) sound incidence is shown in Fig. 13.7. Below f_{c}, the amount of isolation afforded by single-leaf walls is essentially proportional to the mass load, which is proportional to the frequency. This means an increase of 6 dB/oct. Above f_{c}, the stiffness term becomes dominant and proportional to ω^3, amounting to an 18-dB/oct increase. This slope is, however, rarely achieved in praxi, amongst other reasons, due to the way the wall is clamped and to bypasses - refer to Fig. 13.12.

13.4 Sound-Transmission Loss of Double-Leaf Walls

As we just discussed, the sound-transmission loss of single-leaf walls is governed by the mass load. This certainly holds below the coincidence frequency, but it is also usually sufficient above it since the bending stiffness, B', of many wall materials is proportional to the mass load, m''.

In cases where the mass of walls is structurally limited, double-leaf walls – pictured schematically in Fig. 13.8 (b, lower panel) – can be used to achieve sufficient sound isolation.

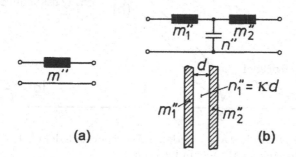

Fig. 13.8. Equivalent circuits for **(a)** single-leaf walls, **(b)** double-leaf walls

In terms of electro-acoustic analogies, a single-leaf wall experiencing perpendicular sound incidence can be represented by a two-port – shown in Fig. 13.8 (a). A double-leaf wall with an air gap inside can be represented as in Fig. 13.8 (b, upper panel). In terms of network theory, the two represent low-pass filters of the 1st and 3rd order, respectively.

With a low-pass filter of the 3rdorder, a transmission loss of 3 times 6 dB/oct or 18 dB/oct can be achieved. This, however, only holds above the fundamental *drum resonance*, ω_0. It may be helpful to compare this configuration to the two-mass resonator discussed in Section 3.7. The drum resonance is determined by the equation

$$\omega_0 = \frac{1}{\sqrt{n''\, m''_{\text{total}}}}\,, \tag{13.25}$$

in which the total effective mass is equal to

$$m''_{\text{total}} = \frac{m''_1\, m''_2}{m''_1 + m''_2}\,. \tag{13.26}$$

Below its fundamental resonance, a double-leaf wall behaves like a single-leaf one because the two leaves are more or less rigidly coupled via the air gap. When the linear dimension of the air gap matches the wavelength of sound waves in air, cavity resonances arise that may reduce the transmission loss. This effect can be reduced by loosely filling the air space between the leaves with absorptive material like mineral wool.

Coincidence effects may occurs in the case of angular sound incidence, but their negative impact can be avoided by making the mass loads of two leaves different.

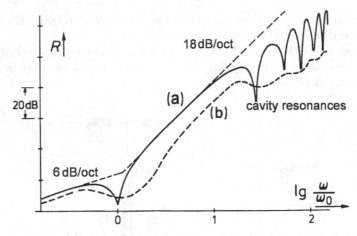

Fig. 13.9. Transmission loss as a function of frequency for a double-leaf wall. (**a**) directed oblique incidence, (**b**) random incidence

Figure 13.9 depicts the principal course of R as a function of frequency for directional and diffuse (random) sound incidences. The cavity resonances are labeled. The advantage of a double-leaf wall over a single-leaf one lies in the region above the drum resonance, where R may increase with a slope of up to 18 dB/oct – at a much lower weight than comparable single-leaf walls.

In real walls in buildings, sound transmission does not only occur via the wall itself but also through the clamping at its rim. Transmission through this path can be substantially reduced by making at least one of the two leaves very compliant for bending waves by setting its coincidence frequency, $f_{c,\,\mathrm{lim}}$, above the spectral region concerned. In this case, bending waves may be transferred into the compliant leaf without being emitted as airborne sound. The bending-wave-compliant leaves may, for instance, consist of gypsum board, metal sheets or heavy foils. A common construction is sketched in Fig. 13.10.

13.5 The Weighted Sound-Reduction Index

In architectural acoustics and related fields the sound-isolation capability of a wall is characterized by an internationally standardized single-number index called the *Weighted Sound-Reduction Index*, R_w. This index is specific to the wall element considered and independent of its actual installation, for instance, in a building.

The procedure for measuring R_w assumes diffuse sound incidence and a relevant spectral region of $100 - 3200$ Hz. Figure 13.11 shows the measurement set-up. The sending room is excited by noise, and the diffuse-field sound-pressure, L_ds, is determined. The receiving room has a known equivalent

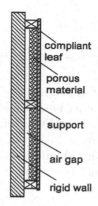

Fig. 13.10. Arming a single-leaf wall with an additional, bending-compliant leaf

absorption area, $A_{\alpha,\,r}$ – refer to Section 12.5 for this quantity. The sound-pressure level, L_{dr}, is measured in the diffuse field of this room.

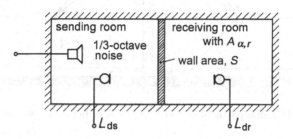

Fig. 13.11. Measurement set-up for the transmission loss of walls, windows etc.

The sound-reduction index, R, is defined as

$$R = 10\lg\left(\frac{I_{ds}}{I_{dr}}\right)\,\mathrm{dB} = 10\lg\left(\frac{P_{ds}}{P_{dr}}\right)\,\mathrm{dB}\,, \tag{13.27}$$

where the sound power impinging on the wall, P_{ds}, is

$$P_{ds} = \frac{S\,W''_{ds}\,c}{4} = \frac{S\,p^2_{ds,\,rms}}{4\varrho c}\,, \tag{13.28}$$

and the sound power transmitted through the wall into the receiving room, P_{dr}, is

$$P_{dr} = \frac{A_{\alpha,\,r}\,W''_{dr}\,c}{4} = \frac{A_{\alpha,\,r}\,p^2_{ds,\,rms}}{4\varrho c}\,. \tag{13.29}$$

These terms may be combined into the ratio,

$$\frac{P_{ds}}{P_{dr}} = \frac{S}{A_{\alpha,\,r}}\,\frac{p^2_{ds,\,rms}}{p^2_{dr,\,rms}}\,, \tag{13.30}$$

and rewritten in logarithmic terms,

$$R = L_{\mathrm{ds}} - L_{\mathrm{dr}} + 10 \lg \left(\frac{S}{A_{\alpha,\mathrm{r}}} \right) \mathrm{dB} \,. \qquad (13.31)$$

Please be aware of the fact that in real installations sound may be transmitted from one room to another via other paths besides the wall itself. Common by-passes are shown in Fig. 13.12. To reduce their effect, the following measures are taken. The flanking walls must be of sufficiently heavy construction, the air gap above a suspended ceiling should be compartmentalized and damped, and the covering floor should not be extended beneath the wall but rather be interrupted.

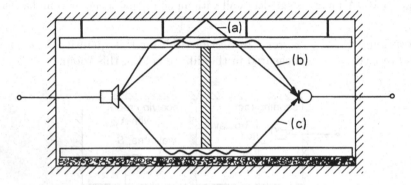

Fig. 13.12. Possible paths of sound transmission between two enclosed spaces

According to the standard, R is measured in $1/3^{\mathrm{rd}}$-oct bands. To obtain a single-number criteria, the frequency curve of the measured R is compared to a reference curve that defines a so-called *Weighted-Sound-Reduction-Index* value of $R_{\mathrm{w}} = 52\,\mathrm{dB}$[1]. The measured curve is then shifted parallel to itself and toward the reference curve in 1-dB steps until the sections below reference remain on average $\leq 2\,\mathrm{dB}$ – depicted in Fig. 13.13. The amount of plus or minus shifting in dB is then added to 52 dB. The resulting Weighted Sound-Reduction Index of the wall is

$$R_{\mathrm{w}} = (52 \pm \mathrm{shifting})\,[\mathrm{dB}] \,. \qquad (13.32)$$

Minimum requirements for R_{w} and R'_{w}, respectively, have been standardized but vary form country to country. Reasonable values are as follows. Walls between apartments should have an $R'_{\mathrm{w}} \geq 52\,\mathrm{dB}$. Between separate dwellings and rooms used for activities that do not belong to the apartment considered, the respective value is 62 dB.

---

[1] When R_{w} has been measured in a test set-up with standardized by-passes, this is indicated by an inverted comma, namely, R'_{w} instead of R_{w}

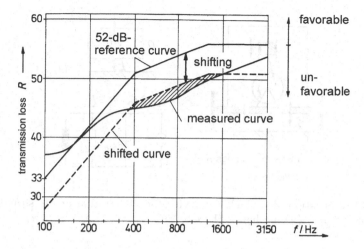

Fig. 13.13. Determination of the Weighted Sound-Reduction Index

13.6 Isolation of Vibrations

Isolation of structure-borne sound often turns out to be difficult in praxi. On the one hand, materials like steel or concrete have only very low damping coefficients for structure-borne sound waves. On the other hand, it is often not possible to construct adequate isolation measures like soft resilient inlays in optimal positions. For these reasons, it is best to take care of preventing structure-borne sound from entering structures like buildings, vehicles, and engines in the first place. This is accomplished by directly isolating the source, usually by providing elastic support. We will elaborate on how elastic support works acoustically.

Since sources of structure-borne sound are usually small compared to the wavelength of structure-borne sound waves, the fundamental principle of vibration isolation can be illustrated with concentrated elements. We further restrict ourselves to an example with only one degree of freedom. If there are more degrees of freedom involved, comparable measures are taken for each of them.

The example with a single-mass vibrator pictured in Fig. 13.14 is known as the *engine-support problem*. The figure shows a sketch of the situation and a mechanic-circuit diagram with an equivalent electric circuit of the 1st-kind analogy.

In our example, it is assumed that the exciting force, $\underline{F}_0$, is a constant sinusoidal force[2]. The task at this point is to tune the system in such a way that the force draining into the ground, $\underline{F}_1$, is minimized. To this end, an *Isolation Index*, R_I, is defined as follows,

---

[2] Alternatively one could, for example, assume constant velocity, $\underline{v}_0$

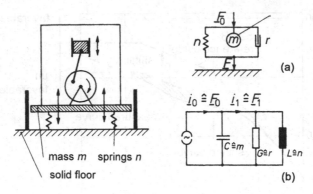

Fig. 13.14. Isolation of a single-mass vibrator. Illustration of the situation and equivalent circuits, **(a)** mechanic, **(b)** electric

$$R_I = 20 \lg \left| \frac{\underline{F}_0}{\underline{F}_1} \right| . \tag{13.33}$$

After a short calculation – based on Fig 13.14 – one gets

$$\frac{\underline{F}_0}{\underline{F}_1} = \frac{r + j\omega m + \frac{1}{j\omega n}}{r + \frac{1}{j\omega n}} . \tag{13.34}$$

From this point, with

$$\omega_0 = \frac{1}{\sqrt{mn}} \quad \text{and} \quad Q = \frac{\omega_0 m}{r} \approx \frac{\omega_0}{\Delta\omega} , \tag{13.35}$$

$$R_I = 20 \lg \sqrt{\frac{\left(\frac{\omega}{\omega_0}\right)^2 + Q^2 \left[\left(\frac{\omega}{\omega_0}\right)^2 - 1\right]^2}{\left(\frac{\omega}{\omega_0}\right)^2 + Q^2}} . \tag{13.36}$$

A plot of this function is given in Fig. 13.15 for various values of the quality factor, Q. Asymptotic values for the slopes of the curves are as follows.

- $Q \to \infty$ ... which is vanishing damping. After dividing (13.36) by Q^2 and neglecting small members of the sums, we get $R_I \simeq 20 \lg \sqrt{(\omega/\omega_0)^4} \sim \omega^2$. This means that the slope approximates $12\,\mathrm{dB/oct}$

- $Q = 1$ ... which is a considerably damped oscillating case. Neglecting small sum members again, we get $R_I \simeq 20 \lg \sqrt{(\omega/\omega_0)^2} \sim \omega$. The slope, then, is $6\,\mathrm{dB/oct}$

Please note that R_I only becomes positive for values of $\omega/\omega_0 > \sqrt{2}$. To isolate a source of structure-borne sound, the support must therefore be tuned to

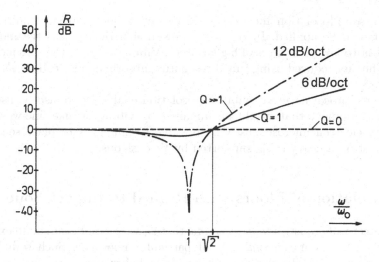

Fig. 13.15. Transmission loss as a function of frequency

a resonance frequency well below the operating frequency, which is usually the frequency of the exciting force. This technique is called *low-end tuning*. Usually one aims for a frequency ratio of $1/5^{\text{th}} - 1/10^{\text{th}}$.

The isolation index decreases with increased damping of the system because the damping acts as a *sound bridge*. Moderate damping is still important, however, because it may limit the *"dancing"* effect experienced by the sound source at or close to the resonance frequency. This may also occur when a rotating engine starts up and runs slowly through the resonance point of the support. It is certainly wise to limit movement at these frequencies with properly adjusted dashpots or even hard mechanic boundaries.

Many types of elastic elements are currently used for acoustic isolation, including steel springs, rubber-metal-compound elements, rubber, and synthetic-foam plates. Progressive spring characteristics like those provided by rubber can help avoid dancing of the source.

If the elastic element in use is known to have a linear characteristic, the resonance frequency, ω_0, can be determined from the amount the element compresses under a load, x_0. This is a handy on-site check for whether the right elastic elements have been installed or if there are accidental sound bridges. The relevant formula is

$$\frac{f_0}{\text{Hz}} \approx \frac{5}{\sqrt{x_0/\text{cm}}}, \qquad (13.37)$$

with

$$\omega_0 = \frac{1}{\sqrt{m\,n}}, \quad n = \frac{x_0}{F}, \quad \text{and applying } F = m \cdot 9.81\,\frac{\text{m}}{\text{s}^2}. \qquad (13.38)$$

When a sound-isolation increase of 12 dB/oct is not sufficient, multi-layer supports can be applied. In terms of transmission theory, this technique is equivalent to low-pass filters of higher order. Vibrations at distinct frequencies can be further reduced using tuned resonating absorbers, also called *vibration extinctors*.

The reciprocal nature of vibration isolation enables it to also be used to protect equipment that would be impaired by vibration, like fine-weighing scales or electron microscopes. One possible technique is to place sensitive equipment on a heavy table supported by air cushions.

13.7 Isolation of Floors with Regard to Impact Sounds

In architectural acoustics *tapping sound*, that is, structure-borne sound excited by walking on floors (*footfall*), is of particular relevance. Such sound can be transmitted to adjacent rooms, especially below the floor, and then be radiated as airborne sound. This also holds for other approximately point-wise sources of excitation like knocking, falling items, pushed chairs, or household appliances.

Isolation against this kind of sound can be accomplished with resilient floor coverings, such as carpets, plastics layers or wooden plates, under-packed with a compliant layer. The achieved isolation can be estimated in the same manner that was described earlier. The quantity used as mass in such an estimation should be that part of the total source mass that is in direct contact with the floor, and the value used as compliance is the one given by the contact area.

The *floating* or *resilient floor* is another very efficient method for reducing tapping sounds. As sketched in Fig. 13.16, such a floor consists of a load-distributing plate, possibly made of cast concrete, supported by a compliant layer, usually made of mineral wool or elastic synthetic foam.

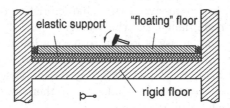

Fig. 13.16. Floating-floor construction for isolation against tapping sounds

The assembly consists of two plates in which bending waves may propagate, the floor itself and the concrete plate on top of it. The latter two are complexly coupled with the elastic under-packing. It has been shown that the bending stiffness of the top plate can be neglected if certain conditions are met. These conditions are as follows. The top plate must be considerably thinner than the floor itself and the under-packing must have a high enough flow resistance to prohibit sound propagation along it.

In most cases the same formulae can be used that have already been derived for isolation with concentrated elements – see Section 12.5. The resonance frequency of the floating floor is thus determined by

$$f_0 = \frac{1}{2\pi} \sqrt{\frac{s''}{m''}} \quad \text{with} \quad s'' = \frac{1}{n''}, \tag{13.39}$$

where m'' is the mass load and s'' the dynamic stiffness, which is the inverse of the dynamic compliance, n''. The dynamic stiffness of a resilient layer is measured according to international standards. It contains the stiffness of the enclosed air, which allows it to account for the fact that isothermal compression dominates within the layer.

In architectural acoustics, the resonance frequency, f_0, should be chosen to be well below $100\,\text{Hz}$, so that human speech is sufficiently isolated from. Extreme care must always be taken to prevent rigid contact between the floating plate with the rigid floor or the walls. Such contact, often called a *sound bridge*, impairs the isolation significantly. One small sound bridge may easily reduce the isolation by more than $10\,\text{dB}$.

14

Noise Control – A Survey

So far we have dealt with sound sources that convert electric energy into mechanic/acoustic energy. These sound sources are mainly used to radiate desired sounds because the mechanic/acoustic signals are easily controlled by the electric ones. In addition to desired sound, there is undesired sound that one would reduce or even eliminate if possible. This kind of sound is called *noise*[1]. A reasonable definition of noise in acoustics must cover various aspects, such as the following one.

> *Noise* is audible sound that disturbs quietness, or the reception of intentional sound, or leads to damage, annoyance or health impairment

In our engineered and industrialized environment, noise has become a significant problem because it tends to be the byproduct or "garbage" of technological processes. Industrial machines and appliances that have not been treated with special noise-control measures radiate $1 - 10\,^0/_{00}$ of their driving power as sound. Although it is only a small fraction of the total power that is converted into sound, considerable sound-power values may still be reached. Table 14.1 lists the sound power of some prototypical sources for comparison.

The sound-power level, L_p, is a temporal and spatial average. In a free sound field it is determined by integrating over all directions of radiation. In a reverberant space the diffuse-field is measured as described in Section 12.6. The sound-pressure level at a certain point in space can be derived from the sound-power level of the source, but that is usually not a simple process.

---

[1] In signal theory the same term is used for stochastic signals that are not prone to carry information

Table 14.1. Sound power of some typical sources

Source type	$\overline{P}$	L_w re 10^{-12} W
Space rocket (Saturn)	$> 10^7$ W	> 190 dB
Jet airplane	10^4 W	160 dB
Large brass orchestra	10 W	130 dB
Large machine tool	1 W	120 dB
Passenger cars on highway	10^{-2} W	100 dB
Normal speech	10^{-5} W	70 dB
Soft whispering	10^{-9} W	30 dB

14.1 Origins of Noise

Reasons why noise is generated are manifold, and a few examples of noise sources are listed below – ordered with respect to their sound excitation types. The list is not intended to be complete.

- Airborne sound sources
  - excitation by explosion or implosion $\longrightarrow$ impulsive sounds
  - excitation by turbulent flow $\longrightarrow$ non-periodic sounds
  - excitation by intermittent flow, e.g., siren, car exhaust $\longrightarrow$ periodic or quasi-periodic sounds
- Structure-borne sound sources
  - excitation by stroke or knock $\longrightarrow$ impulsive sounds
  - excitation by friction $\longrightarrow$ non-periodic sounds
  - excitation by periodic forces, e.g., magnetic or electric ones $\longrightarrow$ periodic or quasi-periodic sounds

In airborne sound sources, forced or free vibrations are excited in air or gas volumes. In structure-borne sound sources, structures are excited by free or forced vibration. Such structural vibrations may be radiated as airborne sound, particularly when large plates like walls, shells or housings are vibrating.

14.2 Radiation of Noise

Radiation of noise follows the general laws of sound radiation that have been discussed throughout this book, especially in Chapters 9, 10 and 11. Radiation has specific directional characteristics that depend on the form of the gas volumes or structures and the type of vibration they experience.

We will now discuss sound radiation by line arrays of non-coherent sound sources and the role of meteorological conditions, two items that have not been considered up to this point. Of course, both items can also be related to the radiation of intentional sound signals, but they are rarely of importance there.

Line Arrays of Incoherent Sources

The significant difference between arrays of coherent and incoherent sound sources is that no interference occurs in the latter case. For this reason, it is not necessary to consider the phase of partial sound pressures when computing the sound field. Partial intensities are added up instead – explained in Section 1.6.

We will consider as an example a line array that is densely occupied by incoherent point sources of 0th order and positioned on a reflecting surface. Such an array may be taken as a model for a heavily used road.

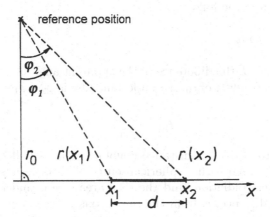

Fig. 14.1. Sound radiation from a line array to a reference position

The line array is schematically plotted in Fig. 14.1. We assume that it experiences a constant load of active sound power, $\overline{P'}$, which is also expressed as constant power per length.

The reference point where the sound field will be computed should be in the far field, where it holds that

$$\frac{p_{\text{rms}}^2}{\varrho\,c} = \left|\vec{I}\right| = I. \tag{14.1}$$

The element, $\mathrm{d}x$, yields a partial intensity at the reference point of

$$\mathrm{d}\left[\int \left|\vec{I}\right| \mathrm{d}\varphi\right]. \tag{14.2}$$

With a spatial angle of 2π since the reflective floor restricts us to only the upper hemisphere, we get

$$\mathrm{d}\left[\int \left|\vec{I}\right| \mathrm{d}\varphi\right] = \frac{\overline{P'}\,\mathrm{d}x}{2\pi\,r^2(x)} = \frac{\overline{P'}\,\mathrm{d}x}{2\pi\,(r_0^2 + x^2)}. \tag{14.3}$$

With the end points of the line array defined to be x_1 and x_2, and the length equal to d, integration yields

$$\underbrace{\int_{\varphi_1}^{\varphi_2} \left| \overrightarrow{I} \right| d\varphi}_{I_\Sigma} = \frac{\overline{P'}}{2\pi} \int_{x_1}^{x_2} \frac{dx}{r_0^2 + x^2}$$

$$= \frac{\overline{P'}}{2\pi} \frac{1}{r_0} \left[\underbrace{\arctan \frac{x_2}{r_0}}_{\varphi_2} - \underbrace{\arctan \frac{x_1}{r_0}}_{\varphi_1} \right]. \tag{14.4}$$

Very long and very short arrays are two especially noteworthy cases. We shall thus discuss them in the following.

• *Very long line array*

In the case of $r_0 \ll d$, the difference of the apparent angles, $\varphi_2 - \varphi_1$, becomes constant and independent of r_0. An important case is $\varphi_2 - \varphi_1 = \pi$, from which it follows that

$$I_\Sigma = \frac{\overline{P'}}{2} \frac{1}{r_0}, \tag{14.5}$$

which amounts to - 3 dB per distance doubling. This is half the attenuation rate of -6 dB associated with a spherical source. This is because the situation imitates cylindrical radiation, and the shell area of a cylinder increases proportional to the distance, r, from the center axis.

• *Very short line array*

Figure 14.2 illustrates that in the case of a very short array, when $r_0 \gg d$, the following approximation applies, with r being the average distance,

$$\tan (\varphi_2 - \varphi_1) \approx \varphi_2 - \varphi_1 \approx \frac{d}{r_0} \cos^2 \varphi. \tag{14.6}$$

Insertion yields

$$I_\Sigma = \frac{\overline{P'} d}{2\pi r^2}, \tag{14.7}$$

which indicates - 6 dB per distance doubling. Very short line arrays may therefore be dealt with as 0th-order spherical sources with a sound power of $\overline{P} = \overline{P'} d$.

Influence of Meteorological Conditions

Temperature and wind profiles are the most relevant meteorological conditions. We shall thus briefly comment on them in the following.

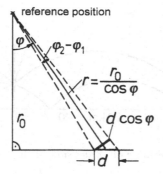

Fig. 14.2. Sound radiation from a very short line array to a reference position

• *Temperature profiles*

The speed of sound, c, is the speed at which a wave front propagates, and it depends on the temperature of the medium according to

$$c \sim \sqrt{T} \sim \sqrt{1/\varrho}. \tag{14.8}$$

Refraction always deflects sound into the colder and thus denser medium – discussed in Section 11.3. Three typical cases are illustrated in Figs. 14.3, 14.4 and 14.5.

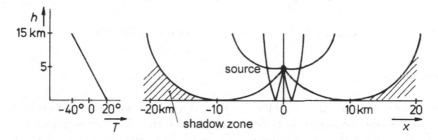

Fig. 14.3. Sound propagation in a normal temperature profile

In Figure 14.3, we see a normal temperature profile that might belong to a sunny day when the ground is heated up. The sound is reflected upwards, creating a shadow zone. Sound from sources beyond an *acoustic horizon* does not come across.

Figure 14.4 shows a case of *inversion* that is characterized by a reversed sign in the gradient of the temperature profile. This situation may occur on a clear summer night when the ground has already radiated its heat while the air is still warm. Sound is able to propagate along very long distances. It is actually under similar conditions that African *drum telephony* becomes effective.

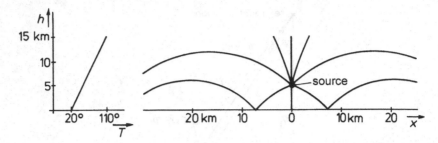

Fig. 14.4. Sound propagation while *inversion* of the temperature profile occurs

Figure 14.5 illustrates the *sound-channel* effect, which may occur in early morning. Such sound channels are common in underwater sound environment and must be considered in SONAR explorations.

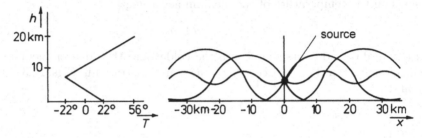

Fig. 14.5. Sound propagation in a *sound channel*

• *Wind profiles*

Due to friction, a wind velocity gradient develops above the surface of the ground. Sound propagation is deflected because the static wind velocity, v_-, superposes with the alternating particle velocity of the sound, $v_\sim$. A shadow zone appears at the windward (luff) side – shown in Fig. 14.6 – but the boundary of the *shadow zone* is blurred due to turbulence[2].

14.3 Noise Reduction as a System Problem

Noise that is generated and radiated from one or more sound sources propagates along one or multiple paths before arriving at one or more receivers (sinks). It seems natural to discuss this situation as a *system* in transmission-theory terms. Measures for noise control in this system require knowledge

---

[2] It is worth noting that the influence of wind leads to violation of the principal of reciprocity

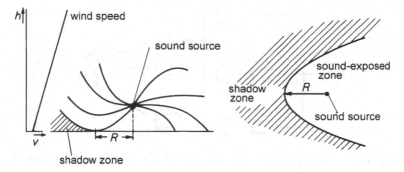

Fig. 14.6. Sound propagation under the influence of wind

of the structure of the system and of the characteristics of its components, including sources, transmission channels and receivers.

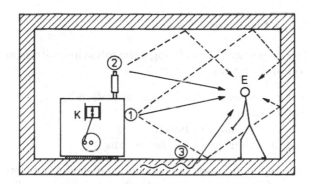

Fig. 14.7. Different kinds of noise radiation and different propagation paths indoors

In Fig. 14.7 the situation is illustrated by a person who is exposed to noise from a rotating engine on an enclosed machine floor. Sound emission happens (1) by airborne sound radiation from vibrating surfaces, (2) by turbulent and intermittent gas flows, and (3) by the vibrating floor. The sound arrives at the exposed person via several different propagation paths – schematized in Fig. 14.8. The planning of noise-control usually includes three groups of tasks, namely, system analysis, goal setting, and decision making. More details are listed below.

System Analysis

The main steps in system analysis are

- Identification of the sources, investigation of the sound-generation mechanisms and determination of sound powers and directional characteristics

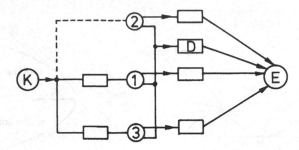

Fig. 14.8. The topology of the system – illustrated for system analysis

- Identification and surveying of the sound-transmission paths. In praxi, it is often useful to explore airborne-sound and structure-borne-sound paths separately
- Investigation of the effect of the noise on the exposed receiver

Receiver-related Limits of Exposure

If the exposed receivers are human beings, the following criteria are important when settling on limiting values.

- Just acceptable interference with communication
- Just acceptable annoyance
- Protection against damage of hearing
- Protection against damage of further organs

Obligatory noise-exposure limits are set by applicable laws, standards and work-protection codes, but it is often difficult to predict the reliability of limits, especially in cases of annoyance due to noise that conveys information.

Adequate Noise-control Measures

When making decisions, it is useful to list all possible measures and weight them according to feasibility and effort. An optimized battery of measures can be compiled on this basis. It is important to also consider extreme measures like replacing machines and completely terminating production in a certain setting. Personal solutions like frequent temporary replacement of persons with extreme noise exposure helps to mitigate harmful effects.

When considering multiple sources and transmission paths, the rules of level addition – introduced in Section 1.6 – require that one always starts with that component that contributes the most to the overall level.

Annoying low level sound signals can be made inaudible by masking them with more pleasant sounds at the same or a slightly higher level. In an open-office space, for example, the noise of the air-conditioning system can be used to mask disturbing voices. Also, the diffuse bubble of voices that is typical for restaurants improves the situation for people at one table by making talk from other tables unintelligible and thus providing a sense of *privacy*.

14.4 Noise Reduction at the Source

It goes without saying that the best method of noise control is avoiding or reducing the generation of noise in the first place. This type of noise control is called *primary* (active) *noise control*, and it starts at the earliest phases of construction. Some relevant aspects are listed and discussed below.

Avoidance/Reduction of Excitation of Airborne Noise

- *Explosions* and *implosions* can usually not be avoided, but their temporal and spectral noise properties may be modified, for example, by redesigning combustion chambers

- *Turbulent flow* creates noise levels that increase with flow velocity with a proportionality constant of up to the 8th power of it. Reduction of flow velocity is therefore of paramount importance and may be accomplished using ducts with larger cross sections and avoiding constrictions and edges. Flow-compliant profiles should be applied instead

- *Intermittent flows* should not be coupled to resonant cavities

Avoidance/Reduction of Excitation of Structure-borne Noise

- *Jolts* can be avoided by constructive measures like making the kinetic sequences steady up to their 3rd derivatives. If this is impossible, the moving masses and their velocities might be reducible. Further relevant precautions include insertion of elastic layers, allowing for some slack in power trains, reduction of slackness where components hit each other, balancing of rotation elements, and provision for phase compensation by tuned vibrators – refer to Section 13.6

- *Friction* can be reduced with lubrication and/or high-quality surfaces. When choosing materials, their acoustic characteristics should definitely be considered. Materials with elasticity and high internal losses are usually preferable

- *Periodic forces* can be minimized by careful balancing, provided that they are mechanically induced. When they are excited by electric or magnetic forces, proper construction is of paramount importance

Often significant progress in sound control can be made by modifying a process. This alternative is thus worthy of keeping in mind.

Avoidance/Reduction of Radiation of Structure-borne as Airborne Noise

Engine elements and housings are to be designed for low airborne sound radiation, among other measures, by minimizing and perforating surfaces. It is also helpful to divide those same surfaces so that knots of vibration result. One can also avoid resonances by adding mass, using plates that are heavy but compliant for bending waves, and/or applying coatings with absorbent layers and/or internal viscous layers for dampening purpose.

14.5 Noise Reduction Along the Propagation Paths

Noise control along propagation paths is called *secondary* (passive) *noise control*. We will now give an overview with respect to air- and structure-borne noise, organized according to the three main noise-reducing effects on the propagation paths, namely, distribution, reflection and absorption.

Reduction by Distribution

During propagation, sound usually spreads to fill out geometrically larger area. As a result, the intensity and sound-pressure level decrease with increasing distance from the source. Airborne noise displays this behavior not only for free-field propagation but also for guided propagation as occurs in branching ducts. Diffraction and scattering may also support distribution by breaking up beams. Geometric distribution is also a factor for structure-borne noise, where it occurs in plates or at the edges and joints of structures. Distribution does not take place in canals or rods with constant or decreasing cross-sectional area or in enclosed spaces beyond the *critical distance* (diffuse-field distance), which is determined by the directional characteristics of the source – refer to Section 12.6. The more the sound is focussed, the longer the critical distance.

The maximum level decrease due to geometric distribution is associated with 0th-order spherical sources in the free field, which experience 6 dB per distance doubling. Only 3 dB per distance doubling can be achieved for very long line arrays and for small sources in shallow rooms because of the geometry of cylindrical wavefronts.

Whenever the situation allows for it, proper geometric placement of sound sources with respect to receivers should be considered as an effective approach to noise control. The farther, the better!

Reduction by Reflection

Reflection can provide isolation from airborne noise by creating shadow zones behind occlusions like walls and barriers. Refraction at boundaries and inhomogeneities also supports isolation as do meteorological conditions. Structureborne sound is reflected at boundaries where the impedance changes. The most effective noise-reduction measure during propagation is reflection, which is implemented by encapsulating the source with non-porous walls and other dividers for airborne sound or with elastic layers for structure-borne sound.

Complete encapsulation[3], however, is often not feasible, especially outdoors or in open-space offices. In these cases one is restricted to noise barriers, which are less effective. The maximum shadowing ability of barriers is reduced by diffraction so that their maximum insertion loss for airborne sound is only about 20 dB.

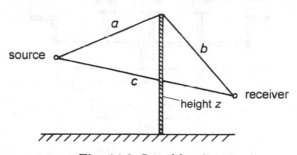

Fig. 14.9. Sound barrier

Diffraction about a barrier can be theoretically described by assuming a line array at the rim of the barrier, but actual calculation of the effect is tedious. For this reason, many noise-control standards provide approximate procedures for estimating the insertion loss. The insertion loss, D_i, increases monotonically with the relative lengths of the detour around the barrier and the wavelength of the sound. An example for an empirical formula corresponding to Fig. 14.9 is

$$D_i = 10 \lg (20.4 \, k_f + 3), \qquad (14.9)$$

where k_f is the *Fresnel number* and equal to

$$k_f = \frac{(a + b) - c}{\lambda/2}. \qquad (14.10)$$

Formula 14.10 requires that the sound source be a 0th-order spherical source, that $k_f > -0.1$, and that the bypass around the edges of the barrier, y, is long enough. A rule is, with z being the barrier height,

---

[3] To increase their efficiency, capsules should have some sound-absorbing material inside

$$\text{either } y > 5z \quad \text{or} \quad y > z + \lambda. \tag{14.11}$$

On the side of the barrier facing the source, a level increase of up to 6 dB may be observed due to reflection from the barrier. This effect can be avoided by equipping the barrier with an absorptive surface[4].

An example of a case where reflection is not viable is given by ducts that carry flowing media, like exhaust systems or air-conditioning ducts. Non-porous walls cannot be applied because the media flow must not be interrupted. Impedance changes, however, may still be implemented in several ways, including coupled, cascaded or branching resonators, detours, bypasses, and steps in the cross-sectional areas – dealt with in Sections 8.5 and 8.6.

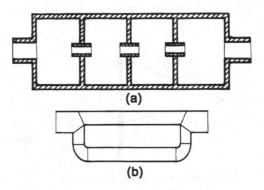

(a)

(b)

Fig. 14.10. (a) Reflection muffler, (b) comb-filter muffler

According to the theory of transmission-line and reactance filters, reflection and interference can be used to build mufflers to spectral specification. See Fig. 14.10 for two examples of how this is built, namely, a low-pass and a comb filter. For the reflection muffler, the different chambers may be of different volume to increase the effective bandwidth. Comb-filter mufflers are delicate since interference minima (comb-filter minima) are narrow and hard to adjusts. Where exactly in a duct system *mufflers* are actually positioned is very important[5].

Reduction by Absorption

Dissipation in the medium is an important reason for absorption of airborne sound, especially at boundaries. The means of absorption for structure-borne sound are dissipation in the materials themselves (internal losses) and absorption due to absorptive coating, padding and inlays.

---

[4] Barriers in enclosed spaces are only sensible if the ceiling above them is absorbing

[5] The sound level may actually increase at the input port of a muffler, due to reflection

When sound propagates outdoors, additional absorption is provided by the ground. The nature of this effect depends on the character of the surface, which may be lawn, bushes, trees or a number of other terrains. This supplementary absorption is often overestimated. As a rule, it is negligible at small distances from the source, $r < 100\,\mathrm{m}$. Beyond that distance, a supplementary loss of $8\,\mathrm{dB}$ per $100\,\mathrm{m}$ is a reasonable estimate for traffic noise propagating over grass and bush or through woodland. This loss is smaller during the winter when the leaves are fallen.

Noise reduction by absorption is often the only reasonable measure to be taken in enclosed spaces with distributed sound sources, like machine-floor halls. For diffuse sound fields we refer to (12.32), according to which

$$p^2_{\mathrm{d,\,rms}} = \frac{4\,\overline{P}\,\varrho\,c}{A_\alpha}\,, \tag{14.12}$$

holds for the sound pressure. This means that a doubling of the equivalent absorbing area results in a $3\,\mathrm{dB}$ decrease of the sound-pressure level. Additional absorption is more effective, the more reverberant the space was initially.

In order to dampen structure-borne sound, panels are either coated with an absorbent *anti-boom* material or provided with an internal absorbent layer, producing what is called a *sandwich panel*.

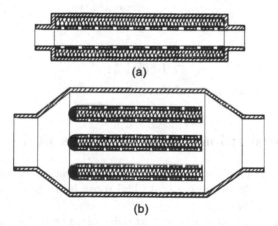

(a)

(b)

Fig. 14.11. Absorption mufflers, (a) duct with absorptive lining, (b) duct with absorptive dividers

Sound propagation in ducts can be abated with *absorption mufflers*. Figure 14.11 shows two examples. The absorptive material is either placed on the walls or in the duct as absorbing dividers[6].

---

[6] If one can accept that media flow will be hindered, the duct may be entirely filled with absorbing (porous) material, a condition called a *throttling muffler*

The sound field in a duct with absorbing walls can be approximated as follows, assuming that the cross-sectional areas are small compared to the wavelengths. The boundaries may be locally-reacting – refer to Section 11.4. If U is the perimeter and A is the cross sectional area, the sound power transmitted into the walls is

$$\mathrm{d}\overline{P}_{\mathrm{wall}}(x) = \frac{p_{\mathrm{rms}}^2(x)\, U\, \mathrm{d}x}{\mathrm{Re}\{\underline{Z}_{\mathrm{wall}}\}} = I_{\mathrm{wall}}(x)\, U\, \mathrm{d}x \,. \tag{14.13}$$

This must be set equal to the incremental loss of axially propagating power, $-I\,\mathrm{d}A$, so that

$$\frac{p_{\mathrm{rms}}^2(x)\, U\, \mathrm{d}x}{\mathrm{Re}\{\underline{Z}_{\mathrm{wall}}\}} + \frac{\mathrm{d}p_{\mathrm{rms}}^2(x)\, A}{\varrho\, c} = 0 \,. \tag{14.14}$$

Rewriting this yields

$$\frac{\mathrm{d}[p_{\mathrm{rms}}^2(x)]}{\mathrm{d}x} + \frac{\varrho\, c}{\mathrm{Re}\{\underline{Z}_{\mathrm{wall}}\}}\, \frac{U}{A}\, p_{\mathrm{rms}}^2(x) = 0 \,, \tag{14.15}$$

which, for the forward progressing wave, has the solution

$$p_{\mathrm{rms}}^2(x) = p_{\mathrm{rms}\,\rightarrow}^2\, \mathrm{e}^{-2\,\check{\alpha}\,x} \,. \tag{14.16}$$

The complex solution for harmonic functions is now approximately

$$\underline{p}_{\rightarrow}(x) = \underline{p}_{\rightarrow}\, \mathrm{e}^{-\check{\alpha}\,x}\mathrm{e}^{-\mathrm{j}\beta\,x} \,, \tag{14.17}$$

with

$$\check{\alpha} = \frac{1}{2}\, \frac{\varrho\, c}{\mathrm{Re}\{\underline{Z}_{\mathrm{wall}}\}}\, \frac{U}{A} \,. \tag{14.18}$$

This is a spatially-damped sound wave.

14.6 Noise Reduction at the Receiver's End

Due to reciprocity, the techniques discussed in this chapter can also be applied to reduce noise on the receiving end. The most important receivers in this context are human beings[7].

For them personal protection against noise often provides the quickest and cheapest noise solution. Unfortunately, there can be problems with acceptance since wearing them may cause discomfort. In the following, we list the most common types of these protectors.

---

[7] Long-time exposure to noise levels above 85 dB in the mid-frequency range may cause permanent impairment of hearing. These levels are not at all uncommon in daily life, especially in places like discotheques or when using headphones. Levels above 130 dB may affect inner organs and may cause symptoms like vertigo and nausea

Personal Hearing Protectors

For insertion losses of 20 – 30 dB, *ear plugs* are a common product. They are usually made from plastic or synthetic foam. The plastic ones, called *oto-plastics*, can be custom-made. The foam ones are compressed by twisting them between the fingers before inserting them into the ear canal, where they expand again to fit.

Ear muffs usually enclose the external ears like circum-aural (closed) headphones. They have sealing rings, sometimes filled with liquid, to tighten against the skull. Insertion losses are 10 – 40 dB, usually increasing with frequency. Ear muffs can be cascaded with ear plugs.

For even higher insertion losses special helmets are available that achieve insertion losses of more than 50 dB at higher frequencies. To protect inner organs, special sound-and vibration-protective suits can be worn, for example, to protect astronauts during the start phase of their space craft.

There are personal hearing protectors that apply so-called *active-noise-control* principles[8]. These devices pick up the noise signals and add them again in inverted phase, after some adequate signal processing. Thus they are able to compensate the noise to a certain extent. Due to technical limitations the active-noise-canceling method is predominantly applicable in a frequency range of up to about 1.5 kHz.

---

[8] In contrast to Section 14.4 the term *active* has a different meaning here. Here it denotes that devices have controlled power sources of their own and an output power that exceeds the input power

15

Appendices

15.1 Complex Notation for Sinusoidal Signals

An arbitrary sinusoidal signal can be written as

$$z(t) = \hat{z}\cos(\omega t + \phi),\qquad(15.1)$$

with the three free parameters amplitude, $\hat{z}$, angular frequency, ω, and (zero)phase angle, ϕ. If the frequency is known and fixed, that is, for a monofrequent signal, only two free parameters are left – amplitude and phase angle.

Arithmetic operations with sinusoidal signals, such as addition, multiplication, differentiation and integration, are complicated in the representation as written above. There is thus a demand for more simple arithmetics to deal with these signals, particularly their amplitudes and phase angles. This can be accomplished in different ways. In this book we use the common symbolic representation of sinusoidal signals by means of so-called *complex amplitudes*.

To this end, the representation above is expanded by a complex imaginary part, but this operation is immediately inverted by forming the real part of the expression again. We thus write

$$z(t) = \mathrm{Re}\left\{\hat{z}\,\cos(\omega_0 t + \phi) + \mathrm{j}\,\hat{z}\,\sin(\omega_0 t + \phi)\right\}.\qquad(15.2)$$

By applying *Euler*'s formula, $\mathrm{e}^{\mathrm{j}\phi} = \cos\phi + \mathrm{j}\sin\phi$, this expression can be rewritten as

$$z(t) = \mathrm{Re}\left\{\hat{z}\,\mathrm{e}^{\mathrm{j}(\omega_0 t + \phi)}\right\} = \mathrm{Re}\left\{\hat{z}\,\mathrm{e}^{\mathrm{j}\phi}\mathrm{e}^{\mathrm{j}\omega_0 t}\right\} = \mathrm{Re}\left\{\underline{z}\,\mathrm{e}^{\mathrm{j}\omega_0 t}\right\}.\qquad(15.3)$$

with the term $\underline{z} = \hat{z}\,e^{j\phi}$ being the complex amplitude as announced above. The original real representation of the sinusoidal signal can be retrieved from the complex amplitude by multiplication with $e^{j\omega_0 t}$ and subsequent forming of the real part, that is, by applying the operator $\mathrm{Re}\{\ldots\}$.

The most relevant rules for calculations with complex amplitudes are given below. For details please refer to the pertinent literature.

- *Addition* and *Substraction*,

$$\underline{z}_1 + \underline{z}_2 = (\hat{z}_1\,\cos\phi_1 + \hat{z}_2\,\cos\phi_2) + j(\hat{z}_1\,\sin\phi_1 + \hat{z}_2\,\sin\phi_2)$$
$$\underline{z}_1 - \underline{z}_2 = (\hat{z}_1\,\cos\phi_1 - \hat{z}_2\,\cos\phi_2) + j(\hat{z}_1\,\sin\phi_1 - \hat{z}_2\,\sin\phi_2) \quad (15.4)$$

- *Multiplication* and *Division*,

$$\underline{z}_1\,\underline{z}_2 = (\hat{z}_1\,\hat{z}_2)\,e^{j\,(\phi_1+\phi_2)}, \qquad \underline{z}_1/\underline{z}_2 = (\hat{z}_1/\hat{z}_2)\,e^{j\,(\phi_1-\phi_2)} \quad (15.5)$$

- *Integration* and *Differentiation*,

$$\int \underline{z}\,dt = \frac{1}{j\omega}\,\underline{z}, \qquad \frac{d\underline{z}}{dt} = j\omega\,\underline{z} \quad (15.6)$$

The rules for integration and differentiation become clear after multiplication of the complex amplitude, $\underline{z}$, with the factor $e^{j\omega_0 t}$, which means reinserting the time dependency.

15.2 Complex Notation for Power and Intensity

Consider a force, $F_z(t)$, exciting a one-dimensional motion of $v_z(t)$ along a path z, for example at the mechanic port of an electro-mechanic transducer. The input energy can be written as follows, whereby from now on the index z is omitted for simplicity,

$$W = \int F(t)\,dz = \int F(t)\,\frac{dz}{dt}\,dt = \int F(t)\,v(t)\,dt. \quad (15.7)$$

The transferred instantaneous power, $P(t)$, is than given by

$$P(t) = \frac{d}{dt}\int F(t)\,v(t)\,dt = F(t)\,v(t). \quad (15.8)$$

Let both the force and the particle velocity be sinusoidal time functions, namely,

$$F(t) = \hat{F}\cos(\omega t + \phi_F) \quad \text{and} \quad v(t) = \hat{v}\cos(\omega t + \phi_v). \quad (15.9)$$

This leads to the expression

$$P(t) = \hat{F}\cos(\omega t + \phi_F)\,\hat{v}\cos(\omega t + \phi_v). \quad (15.10)$$

Execution of the multiplication with $\cos\alpha\,\cos\beta = \frac{1}{2}[\cos(\alpha+\beta)+\cos(\alpha-\beta)]$ renders

$$P(t) = \underbrace{\frac{\hat{F}\,\hat{v}}{2}\,[\cos(2\omega t + \phi_{\mathrm{F}} + \phi_v)]}_{\text{alternating}} + \underbrace{\frac{\hat{F}\,\hat{v}}{2}\,[\cos(\phi_{\mathrm{F}} - \phi_v)]}_{\text{constant}}. \qquad (15.11)$$

When determining the time average of the transmitted power, $\overline{P}$, the first part, which is alternating with double frequency, does not contribute. The average power is solely given by the second part, namely,

$$\overline{P} = \frac{\hat{F}\,\hat{v}}{2}\,[\cos(\phi_{\mathrm{F}} - \phi_v)]. \qquad (15.12)$$

This average power is also called *active power* or *resistive power*.

For a complex notation with the complex amplitudes of the force and the particle velocity, that is

$$\underline{F} = \hat{F}\,e^{j\phi_{\mathrm{F}}} \quad \text{and} \quad \underline{v} = \hat{v}\,e^{j\phi_v}, \qquad (15.13)$$

we try an approach that leads us to a *complex power*, $\underline{P}$, as follows,

$$\underline{P} = \frac{1}{2}\,[\underline{F}\,\underline{v}*] \quad \text{with} \quad \underline{v}* = \hat{v}\,e^{j(-\phi_v)}, \qquad (15.14)$$

where the term $\underline{v}^*$ is called the complex conjugate of $\underline{v}$. Some elaboration finally results in

$$\underline{P} = \overline{P} + j\,Q = \underbrace{\frac{\hat{F}\,\hat{v}}{2}\cos(\phi_{\mathrm{F}} - \phi_v)}_{\text{active power}} + j\,\underbrace{\frac{\hat{F}\,\hat{v}}{2}\sin(\phi_{\mathrm{F}} - \phi_v)}_{\text{reactive power}}. \qquad (15.15)$$

The real part of $\underline{P}$ is the active power, $\overline{P}$, as noted above. The imaginary part, Q, is called *reactive power* and has no direct physical relevance. Please note that by taking the complex conjugate of the particle velocity, v^*, in (15.14) and not that of the force, we have chosen that the reactive power of mass is counted positive.

What holds for the power, also holds for the intensity, which is power per area. *Complex intensity* thus results as

$$\underline{I} = \frac{1}{2}\,[\underline{p}\,\underline{v}*]. \qquad (15.16)$$

Consequently, we denote the *active intensity*, $\mathrm{Re}\,\{\underline{I}\}$ also as $\overline{I}$ in this book.

15.3 Supplementary Textbooks for Self Study

The current textbook suffices as the sole teaching material for an introductory course given by experienced academic teachers who are able to provide specific explanations and stress relevant topics according to the prior knowledge and special interests of their students.

Our book is also suitable for self study. In this case, however, we suggest that the reader may use further textbooks in parallel. Some suggestions are given in the following. The list only contains books in the English language and does not claim to be complete.

- Kuttruff, H. (2007), *Acoustics – an introduction*, Taylor & Francis, London–New York

- Raichel D.R. (2006), *The science and applications of acoustics*, 2nd ed. Springer, Berlin–Heidelberg–New York

- Finch, R.D. (2005), *Introduction to acoustics*, Pearson Prentice Hall, Upper Saddle River, New Jersey

- Möser, M. (2004), *Engineering acoustics – an introduction to noise control*, Springer, Berlin–Heidelberg–New York

- Rossing, Th., Moore, F.R., & Wheeler, P.A. (2002), *The science of sound*, 3rd ed. Addison–Wesley, New York, etc.

- Kinsley, L.E., Frey, A.R., Coppens, A.B. & Sanders, J.V. (2000), *Fundamentals of acoustics*, 4th ed., John Wiley & Sons, Hoboken, New York

- Blackstock, D.T. (2000), *Fundamentals of physical acoustics*, John Wiley & Sons, Hoboken, New York

- Pierce, A.D. (1981), *Acoustics – an introduction to its physical principles and applications*, McGraw Hill, New York

- Ford, R.D. (1970), *Introduction to acoustics*, Elsevier, Amsterdam–London–New York

- Randall, R.H. (1951), *An introduction to acoustics*, Reprint 2005, Dover Publications, Mineola, New York

15.4 Exercises

Note: Solutions to the problems will be distributed on a peer-to-peer basis via <http://symphony.arch.rpi.edu/~xiangn/SpringerBook.html>. In case of failure refer to the home page of the book, to be found under http://www.springer.com/engineering/signals/book/978-3-642-03392-6.

Problems to Chapter 1 – Recapitulation of Complex Notation

Problem 1.1. Derive how sinusoidal oscillation quantities can be written in complex complex notation.

Problem 1.2. Given a sinusoidal voltage as a function of time, $u(t) = \hat{u} \cos(\omega t + \phi)$, find $i(t)$ in Fig. 15.1 by applying complex notation.

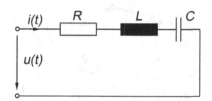

Fig. 15.1. Serial circuit of R, L and C

Problem 1.3. Given a reference pressure of $p_{0,\,rms} = 20\,\mu\mathrm{N/m}^2$.

(a) Find the sound-pressure level of $p_{rms} = 0.1\,\mathrm{N/m}^2$ and $p_{rms} = 10^2\,\mathrm{N/m}^2$, respectively

(b) Evaluate the sound pressure levels for the case that the sound pressures under (a) are three-times larger

Problem 1.4. Which frequency ratios and which logarithmic frequency intervals are represented by three semitones and by twelve semitones?

Problem 1.5. Establish a table for the addition of sound-pressure levels of multiple incoherent sound sources of equal level and equal distance to the receiving point. The number of sound sources is 1, 2, 3, 4, 5, 8, or 10.

Problem 1.6. In an open-space office there are multiple noise sources (talking persons) spatially distributed as shown in the sketch Fig. 15.2. Among them, sources 1–6 are located along a circle of radius a around the receiver, E, with a mutual distance of a between them. Source 7–18 are located apart from each other by, again, a distance of a and along a circle of radius $2\,a$ around the receiver, E. The sound pressure level is assumed to decrease by $5\,\mathrm{dB}$ when doubling the distance in this space. Assuming that the sound pressure level is $65\,\mathrm{dB(A)}$ at a 1-m distance from any single source (talker) for all directions, find the sound pressure level at receiver, E, as a function of the number of sound sources ($a = 4\,\mathrm{m}$).

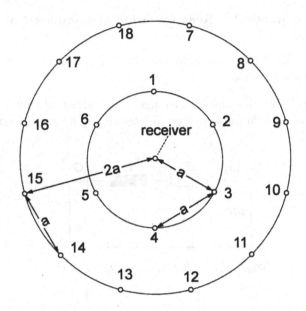

Fig. 15.2. Multiple sound sources of equal intensity in an open-space office

Problems to Chapter 2 – Differential Equations, Free and Forced Oscillation, Resonance Curves, Complex Power in Mechanic Systems

Problem 2.1. Given a mechanic parallel oscillator with mass, m, compliance, n, and damping, r,

(a) Establish a corresponding differential equation for harmonic-force excitation, $F_0(t) = \hat{F}_0 \cos(\omega t)$, in a suitable variable

(b) Find a solution of the differential equation if there is no external excitation

(c) Find the possible types of solutions of the differential equation for excitation as given in (a)

Problem 2.2. A mechanic parallel oscillator be excited by a sinusoidal force of constant amplitude, $F_0 = $ const, which is independent from frequency.

(a) Calculate and graph the trajectory of the velocity, v, in the complex plane as function of frequency, f

(b) Find the relationship between the damping coefficient, $\delta = r/2\,m$, and the bandwidth, $\Delta\omega$, as defined by the half-value (-3 dB) on the resonance curve

(c) Give a justification of the following statement as can be applied for the determination of the quality factor, Q, by inspection of an oscilloscopic plot.

"The quality factor is approximately represented by the number of oscillations of a damped oscillation that precede a 96 % reduction of the starting amplitude"

Problem 2.3. Discuss why the following definition of complex power in a mechanic oscillator system makes sense,

$$P = \frac{1}{2}\,F\,v^* \,.$$

Problem 2.4. Given a lossy electric series resonator.

(a) Determine the following quantities using a graph in the complex plane

Input impedance	$\dots Z$
Input admittance	$\dots Y$
Resonance frequency	$\dots \omega_0$
Characteristic resistance	$\dots Z_0$
Quality factor	$\dots Q$

(b) Derive the relation between bandwidth, $\Delta\omega$, and quality factor, Q

(c) Graph the magnitude of the input impedance, Y, in double-logarithmic representation as a function of frequency

Problems to Chapter 3 – Analogous Electrical Circuits, Acceleration and Velocity Measurement

Problem 3.1. For the mechanic system shown in Fig. 15.3 construct an analogous electric circuit using analogy #2, that is, $F \hat{=} i$ and $v \hat{=} u$.

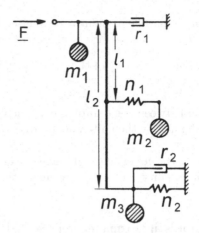

Fig. 15.3. Mechanic circuit to be converted into an equivalent electric one by analogy #2

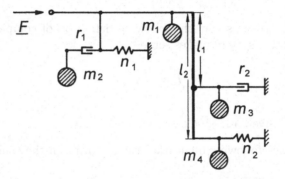

Fig. 15.4. Mechanic circuit to be converted into an equivalent electric one by analogy #1

Problem 3.2.   For the mechanic system shown in Fig. 15.4, construct an analogous electric network using analogy #1, that is, $\underline{F} \hat{=} \underline{u}$ and $\underline{v} \hat{=} \underline{i}$.

Problem 3.3.   A mechanic two-mass oscillator – see Fig. 15.5 – is to be used as a tool to determine displacement, ξ, velocity, $\underline{v}$, and acceleration, $\underline{a}$, of an oscillating mass, m. To this end, displacement, $\underline{\xi}_0$, and velocity, $\underline{v}_0$, are measured between the two masses m_0 and m_1.

(a) Derive the functions $\underline{\xi}_0 = f(\underline{v})$ and $\underline{v}_0 = f(\underline{v})$

(b) Which oscillating quantities of mass m can be determined in measuring
   $\underline{\xi}_0$ and $\underline{v}_0$,

   – For $\omega \gg \omega_0$ (low-end tuning)?
   – For $\omega \ll \omega_0$ (high-end tuning)?

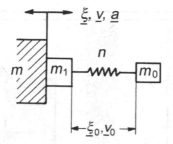

Fig. 15.5. Two-mass arrangement for vibration measurements

Problem 3.4. For the acoustic system shown in Fig. 15.6, construct an analogous electric circuit using lumped elements.

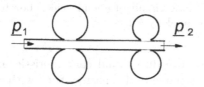

Fig. 15.6. Acoustic circuit to be converted into an equivalent electric one by appropriate analogy

(a) Determine the transfer function $\underline{H}(\omega)$ and graph the course of

$$|\underline{H}(\omega)| = \frac{|\underline{p}(\omega)_2|}{|\underline{p}(\omega)_1|}$$

(b) What are the properties of this network in terms of frequency response? How can one flatten the frequency curve?

Problem 3.5. Construct a mechanic-acoustic network using analogy #2 for the following loudspeaker box – Fig. 15.7. The oscillating coil is excited by a constant force, $\underline{F} = $ constant. Neglect the volume behind the coil and the radiation impedance.

Problems to Chapter 4 – Electromechanic and Electroacoustic Transduction

Problem 4.1. Derive the directional characteristics for arbitrary linear combination of two collocated sound sources, one with spherical and one with figure-of-eight characteristic. Sketch the following special cases: sphere, wide-angle cardiod, cardiod, super cardiod, figure-of-eight.

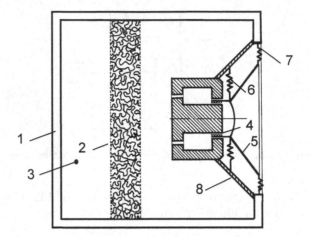

Fig. 15.7. Mechanic-acoustic circuit of a loudspeaker box to be converted into an equivalent electric one by analogy #2

Problem 4.2. Sketch the directional characteristic of a line microphone, that is, a microphone at the end of a narrow tube with slit opening.

Problem 4.3. Pressure-gradient receivers are implemented in praxi as pressure-difference receivers. Determine the magnitude as a function of frequency for such a microphones, given the distance, d, to the measurement point.

Problem 4.4. In a set-up to measure the sensitivity coefficient of a structure-borne-sound transducer, the transducer under test is rigidly connected to two other transducers, one of which is reversible. Mass and compliance of the rigid connection elements are given. Determine the sensitivity coefficient of the transducer under test, based solely on the electric measurements where the reversible transducer is used once as transmitter, and then as receiver.

Problem 4.5. Transmission systems are often mathematically represented by chain matrices.

(a) What advantages does the chain-matrix representation have?

(b) What do the matrix elements mean?

(c) Pure acoustic or electric networks for real transducers can be presented in a form as shown in Fig. 15.8. Find corresponding matrix elements

Problem 4.6. Determine the directional characteristic of a microphone as schematically depicted in the Fig. 15.9.

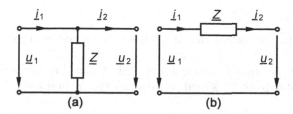

Fig. 15.8. Two-port representation of acoustic and/or electric elements

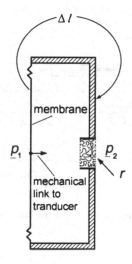

Fig. 15.9. Schematic plot of a microphone with rear opening. $\underline{p}_1$, $\underline{p}_2$ ...sound-pressure signals at front- and backside

Problems to Chapter 5 – Magnetic-Field Transducers

Problem 5.1. Establish an equivalent circuit for an electromagnetic sound transducer that considers the inductance of the arrangement as well as the field compliance.

Problem 5.2. Given a moving-coil microphone as depicted in Fig. 15.10 and with the following parameters.

m_m ... Mass of membrane and coil
n_n ... Compliance of the support
r_m ... Friction of the support
m_1 ... Mass of co-vibrating air
r_1 ... Friction in the air, particularly in the air gap
m_2 ... Acoustic mass of the air in duct 2
r_2 ... Friction due to porous material in duct 2
n_2 ... Compliance of the air in chamber 2

(a) Establish the equivalent circuit for the moving-coil microphone

(b) Discuss the frequency course (curve) of the membrane velocity given a frequency-independent sound pressure on the membrane

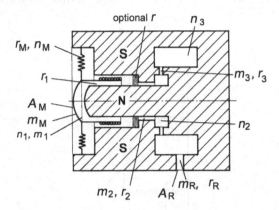

Fig. 15.10. Schematic plot of a moving-coil microphone.

Problems of Chapter 6 – Electric-Field Transducers

Problem 6.1. The following data are given for a dielectric microphone.

Membrane area	$\ldots A = 9.1 \cdot 10^{-4}\,\mathrm{m^2}$
Membrane thickness	$\ldots h = 10^{-5}\,\mathrm{m}$
Membrane distance	$\ldots x_0 = 2 \cdot 10^{-5}\,\mathrm{m}$
Membrane-material density	$\ldots \rho_m = 8 \cdot 10^3\,\mathrm{kg/m^3}$
Membrane-resonance frequency	$\ldots f_0 = 12\,\mathrm{kHz}$
Polarization voltage	$\ldots u_- = 150\,\mathrm{V}$
Resistance	$\ldots R_r = 10\,\mathrm{M\Omega}$

It is assumed that only half of the membrane mass actually moves. Dielectric loss and membrane loss should not be considered.

(a) Determine the elements of the simplified electric equivalent circuit, consider the field compliance

(b) Determine the frequency course (curve) of the microphone

(c) What is the transducer coefficient, u/p, within its effective frequency range?

Problem 6.2. Given an electret microphone as plotted in Fig. 15.11, whereby

    1 ... thin metal layer with a conductivity $\to \infty$
    2 ... isolating layer
    3 ... electret with surface polarization, σ
    4 ... air gap,
    5 ... metallic back plate with a conductivity $\to \infty$

In this microphone, the air gap, d_4, varies as

$$d_4 = d_{40} + d_{41} \sin(\omega t) \ \ \text{with } d_{41} \ll d_{40}.$$

(a) Calculate approximately the AC component, $u_\sim$, of the voltage, u, on an electret microphone

(b) What is the AC component of the voltage, u, on a dielectric microphone of the same construction, driven by a DC voltage, u_{0-}, but without electret? Find the equivalent DC voltage, u_{0-}, for the electret microphone with $\sigma = 10^{-8}\,\mathrm{C/cm^2}$, $d_3 = 1.3 \cdot 10^{-3}\,\mathrm{cm}$ and $\varepsilon_2 = \varepsilon_3 = 3\,\varepsilon_0$

c) What is transducer coefficient, u/p, of an electret microphone below its resonance frequency?

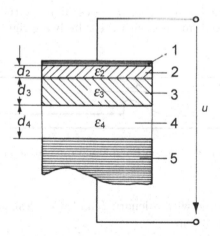

Fig. 15.11. Schematic plot of an electret microphone

Problem 6.3. A quartz as used in an oscillator possesses $R_1 = 60\,\Omega$, $L_1 = 1.5\,\mathrm{H}$, $C_2 = 0.016\,\mathrm{pF}$, and $C_3 = 60\,\mathrm{pF}$ – see Fig. 15.12.

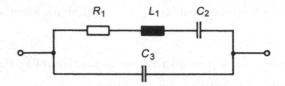

Fig. 15.12. Equivalent circuit of a quartz oscillator

(a) Determine both the parallel and the series resonance, the frequency interval between the two resonances, and the quality factor. Sketch the admittance in the complex plane

(b) How does an additional capacitor, C, parallel to the quartz influence these quantities?

Problems to Chapter 7 – The Wave Equation, Propagation in Tubes

Problem 7.1. Given the one-dimensional acoustic wave equation in lossless fluid.

(a) Show that $d'Alembert$'s solution fulfills this equation

(b) A sawtooth-formed impulse propagates along a tube – see Fig. 15.13. Sketch the reflected impulse with the tube being rigidly terminated

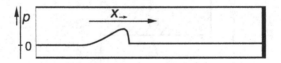

Fig. 15.13. A form-of-a-sawtooth impulse, propagating along a narrow tube

Problem 7.2. A tube with a length L and wave resistance of Z_L is terminated by an impedance Z_0.

(a) Determine, in general form, the input impedance of the tube

(b) Consider the special cases of $Z_0 = 0$, $Z_0 = \infty$, and $Z_0 = Z_w$. What is the sound pressure distribution for these special cases?

Problem 7.3. *Kundt*'s tube is used for measuring arbitrary acoustic impedances. Derive the relations between locations of the standing-wave nodes – and anti-nodes – and the termination impedance.

Problem 7.4. Derive the wave equation of an oscillating string.

Problem 7.5. Determine the eigen-modes (resonances) of

(a) an open organ pipe of 2 m length

(b) a closed (stopped) organ pipe of 2 m length

(c) a rectangular room with linear dimensions of 5 m x 6 m x 10 m

Problems to Chapter 8 – Horns and Ducts

Problem 8.1. A horn radiator is made of an electrodynamic loudspeaker, a pressure chamber and an exponential horn – Fig. 15.14. The horn is assumed long enough, so that reflected waves from the opening end can be ignored. Furthermore, the pressure chamber is very small in comparison with the shortest wave lengths under consideration.

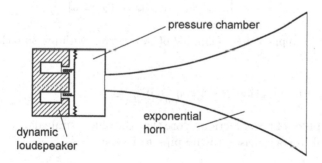

Fig. 15.14. Horn loudspeaker with pressure chamber

(a) Determine the mechanic input impedance of the horn for $\omega \gg \omega_g$. The horn be excited by a constant velocity. Find the horn's flare coefficient so that the radiated power, above a lower limiting frequency of $f_u = 500\,\text{Hz}$, does not vary by more than 3 dB.

(b) Develop an equivalent electric circuit for the pressure chamber

(c) Develop an equivalent circuit for the entire system with $\omega \gg \omega_g$. Simplify the equivalent circuit for the frequency range where frequency course remains constant

(d) Determine the transformation coefficient (area ratio) of the pressure chamber so that the horn radiates the maximum sound power, whereby

Specific impedance of air $\ldots \rho c = 406 \,\mathrm{Ns/m^3}$
Input resistance of the loud speaker $\ldots R_i = 24 \,\Omega$
Inner resistance of the signal generator $\ldots R_g = 24 \,\Omega$
Square value of the transducer coefficient $\ldots B^2 l^2 = 44 \,\mathrm{Vs/m^2}$
Mechanic resistance $\ldots r = 1.2 \,\mathrm{Ns/m}$
Diameter of the speaker membrane $\ldots d = 6 \,\mathrm{cm}$

Problem 8.2. Given an arrangement as sketched in Fig. 15.15.

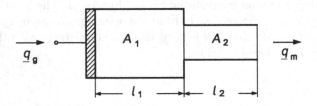

Fig. 15.15. Stepped tube composed of two sections with pressure chamber

(a) Determine the function $\underline{H}(\omega) = \underline{q}_m(\omega)/\underline{q}_g(\omega)$

(b) Find the poles of this function. Note that the output impedance is assumed very small in comparison to the pipe resistance

Problems to Chapter 9 – Sound Sources

Problem 9.1. Given a sound source of 0th–order in free field (a breathing sphere) with a radius of a. The sound velocity on its surface is $\underline{q}_a$

(a) Determine the pressure and velocity distribution in the space outside the sphere

(b) Determine the source strength, $\underline{q}_0$, of the point source which generates the same sound field

Problem 9.2.   Determine and compare the radiated power of a spherical sound source of 0th and 1st order, respectively. Find the dependency of sound intensity on the distance to the source.

Problem 9.3.   An elementary sound source of 0th order is positioned at the origin of an infinitely long conical horn with an opening angle of Ω. The source is mounted in such a way that all its volume velocity is released into the horn. Compare sound pressure, velocity, power, and intensity in the horn with those of the same source in a free field.

Problem 9.4. Given a dipole with the dipole momentum μ_d.

(a) Sketch the directional characteristic of this dipole

(b) An additional 0th–order point source, $\underline{q}_0$, is brought to the same position as that of the the dipole. Find the magnitude and phase of $\underline{q}_0$ so that, in the far field, the entire set-up possesses the following directional characteristics,
$$\Gamma = \frac{1 + \cos\theta}{2}$$

(c) Sketch this directional characteristic. What kind of directional characteristic does it represent?

Problem 9.5. Given a line array with a source-strength load (volume velocity per length) of
$$\underline{q}'(x) = \underline{q}'_0\, e^{-x^2/2h^2}.$$

(a) Determine and sketch the directional characteristic under the assumption that the effect of finite length of the line array is negligible

(b) How does the directional characteristic change if the volume-velocity-load of (a) is modified by the following phase function,

$$\phi = (\beta\ \cos\theta_1)\, x$$

Problems to Chapter 10 – Piston Membranes, Diffraction and Scattering

Problem 10.1.   Using *Fraunhofer*'s approximation of the *Huygens–Fresnel* integral, show that the directional characteristic of an oscillating plane area is expressed by
$$\Gamma(\theta,\phi) = \Gamma(\theta)\,\Gamma(\phi),$$
if $v(x,y)$ can be separated as follows,

$$v(x,y) = v_0 \, \psi_{x(x)} \, \psi_{y(y)},$$

where $\psi_{x(x)}$ and $\psi_{y(y)}$ are weighting function of unity dimension.

Problem 10.2. Find the directional characteristic of a line array of length $2h$ along the x-direction. The line source consists of elementary sources with cardioid directional characteristic along the y-direction.

Problem 10.3. Calculate the far-field directional characteristic of a circular piston membrane.

Problem 10.4. A plane wave hits a quadratic opening of a plane and rigid wall of infinite size. The edge length of the opening is $a = 6.7\,\lambda$. Perform *Huygens–Fresnel*'s zone construction for a reference point in a distance of $r_0 = 3\,\lambda$ perpendicular to the center of the opening.

(a) Sketch the zones from the reference point

(b) Determine the zones according to the order

(c) Show that the pressure in the reference point consists of a sum of undisturbed zone parts and edge refractions at the opening

Problems to Chapter 11 – Dissipation, Reflection, Refraction, and Absorption

Problem 11.1. A *Kundt*'s tube be considered lossy, that is, with a complex propagation coefficient, $\underline{\gamma} = \alpha + j\,\beta$. How would the losses influence the standing waves – and therefore the measurement results?

Problem 11.2. A sound wave is incident with an angle of θ_1 upon a boundary between two different media, Z_{w_1}, β_1 and Z_{w_2}, β_2. Calculate the reflected and transmitted waves. Express the reflection factor and absorption coefficient. What conditions will lead to optimal absorption?

Problem 11.3. Discuss the dependence of the degree of absorption of a wall on the ratio $\underline{Z}_{\text{wall}}/Z_{\text{w, air}}$.

Problems to Chapter 12 – Geometric Acoustics

Problem 12.1. Design a reasonable room boundary for an auditorium as sketched in Fig. 15.16.

Problem 12.2. A rectangular room with linear dimensions of $l_x = 25\,\text{m}$, $l_y = 16\,\text{m}$ and $l_z = 6\,\text{m}$ is given. The absorption coefficients for the floor, the ceiling and the walls are $\bar{\alpha}_1 = 0.2$, $\bar{\alpha}_2 = 0.7$ and $\bar{\alpha}_3 = 0.4$, respectively.

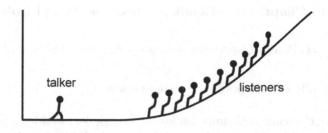

Fig. 15.16. Part of the section of an auditorium – with a talker position and a listeners' area

(a) Calculate the reverberation time of the unoccupied room

(b) How does the reverberation time change when the room is occupied by 150 people? Assume an equivalent absorption area of $0.5\,\mathrm{m^2}$ / person)

(c) Find the critical distance of a dipole source for this room

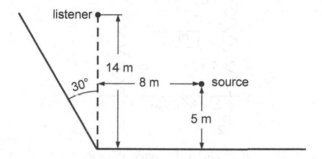

Fig. 15.17. A sound source and a receiver in front of an oblique wall

Problem 12.3. What a boundary profile is particularly suitable for constructing a "whisper gallery"?

Problem 12.4. Given a sound source and a receiver in front of an oblique (non-rectangular) corner – see Fig. 15.17.

(a) Determine the mirror images of 1st and 2nd order for the sound source S

(b) Estimate the impulse sequences originating from the above mentioned mirror-image sources at the receiver position when the source sends out a short impulse. Let the absorption coefficient be $\bar{\alpha} = 0.5$ – propagation losses in air be negligible

Problems to Chapter 13 – Bending Waves and Sound Isolation

Problem 13.1. Bending waves are propagating on a single-leaf wall of infinite size.

(a) Derive a differential equation for bending waves

(b) Calculate the sound-transmission loss of a single-leaf wall for a plane wave of oblique-incident by applying the above-derived differential equation

Problem 13.2. Develop an equivalent electric circuit for a double-leaf wall and determine the transmission loss. How does the transmission loss behave above main resonance, the so-called *drum resonance?*

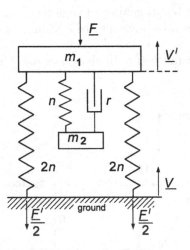

Fig. 15.18. Dynamic absorber

Problem 13.3. To isolate a foundation against vibrations of mass M, a dynamic absorber according to Fig. 15.18 is used. Calculate the transfer ratio F/F' when exciting the mass. Also calculate the ratio v/v'.

Problem 13.4. A bending wave, $\underline{v}_z(y) = \hat{v}\,e^{j\,\beta_b\,y}$, propagates on a sheet of infinite extension along y-direction. The width of the sheet is l. Determine the sound radiated from the sheet at a large distances from the sheet. What kind of directional characteristic is shown by the radiation?

Problems to Chapter 14 – Noise reduction

Problem 14.1. Find a formula for the sound intensity level of

(a) A very long incoherent line source

(b) A very short incoherent line source

Problem 14.2.   Consider a noise-barrier arrangement as depicted in Fig. 15.19.

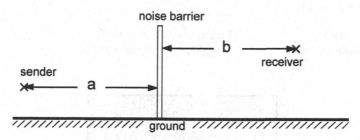

noise barrier

Fig. 15.19. Noise barrier

(a) For $a, b \gg h_{\text{barrier}} - \min(h_{\text{sender}}, h_{\text{receiver}})$, find an adequate approximation formula for the insertion loss, D_i, for example, from applicable standards

(b) Find the insertion loss, D_i, of the barrier for the case of

$a = 14\,\text{m}$ ...4–lane road
$b = 20\,\text{m}$
$h_{\text{sender}} = 0.6\,\text{m}$ ...average height of passenger cars
$h_{\text{receiver}} = 2\,\text{m}$ ...average height of 1st-floor windows
$h_{\text{barrier}} = 4\,\text{m}$
$f = 500\,\text{Hz}$

(c) Find a formula to express the variation of D_i in dB when increasing the frequency by a factor of two or even more. Is this formula valid for averaged levels as well?

Problem 14.3.   Consider the noise-protection arrangement of Fig. 15.19 for the following case. The source is an infinitely long line-array and the noise barrier extends parallel to this line source.

(a) Calculate the insertion loss D_i of this arrangement.

(b) Calculate the insertion loss D_i for the data listed under Problem 14.2 (b) and compare both results

Problem 14.4. The wall of a tube – see Fig. 15.20 – is covered by porous material, namely, either cotton or mineral wool. The following material specification may apply.

For cotton $\ldots \Xi = 1\,\mathrm{g\,cm^{-3}s^{-10}}$, $\sigma = 0.95$
For mineral wool $\ldots \Xi = 100\,\mathrm{g\,cm^{-3}s^{-10}}$, $\sigma = 0.755$

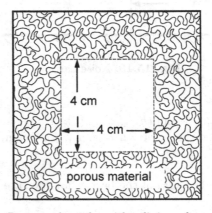

4 cm

4 cm

porous material

Fig. 15.20. Rectangular tube with a lining of porous material

Calculate the sound damping in the tube for the following condition.

- The cross-section of the remaining space in the tube is small in comparison with the wave lengths such that the sound pressure across the sectional area can be considered constant
- Any sound reflected from the rigid tube wall and thus passed twice through the porous material is considered negligible
- The frequency of the sound to be damped is 1 kHz. How thick must the porous-material lining be to fulfil the 2nd condition as to which the sound should at least be attenuated by 20 dB when passing the porous material twice?

Problem 14.5. Refraction is described by *Snellius'* law (*Snell's* law),

$$\frac{c}{\sin \theta} = c_{\text{trace}} .$$

(a) On the basis of this law, derive a general integral-equation for the trajectory of sound rays, $x = \int f(z)$, for the case that the sound velocity is a sole function of height, that is, $c = c(z) \neq c(x, y)$. As sketched in Fig. 15.21, the source is in the position $\{x_0, z_0\}$. The sound ray has an initial slope of

$$\frac{dx}{dz}(z_0) = \tan \Theta.$$

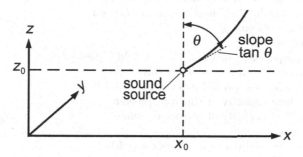

Fig. 15.21. Refraction of sound rays

(b) Solve the integral-equation for the praxis-relevant case of
$c(z) = c_0 - \xi z$ with $\xi \geq 0$ or $\xi \leq 0$, and $x_0 = 0$

(c) Calculate the acoustic horizon for $\xi > 0$, that is, the distance from the source from which the source on the ground can no longer be heard. Use the following data.

$\xi = 1 \cdot 10^{-2}/\mathrm{s}$
$c_0 = 343.6\,\mathrm{m/s}$
$z_0 = 2\,\mathrm{m}$

15.5 Letter Symbols, Notations and Units

Roman-Letter Symbols

a	Acceleration
A	Area
A_α	Equivalent-absorption area
$A_\perp$	Effective area (area perpendicular to particle velocity)
b	Breadth, width, also: substitution for $\beta d \cos\delta$
B	Magnetic-flux density
B'	Bending stiffness
c	Sound-propagation speed in the free field (speed of sound)
c_b	Propagation speed of a bending wave
c_ϵ	Sound-propagation speed in an exponential horn
c_p	Specific heat capacity at constant pressure
c_v	Specific heat capacity at constant volume
C	Capacitance
C'	Capacitance load (capacitance per length)
D	Dielectric-displacement density
D_i	Insertion loss
e	Piezoelectric coefficient
E	Electric field strength
f	Frequency
f_c	Coincidence frequency
F	Force
$\underline{g}$	Product of sound pressure and radius, $\underline{g} = \underline{p}\, r$
G	Electric conductance
G'	Conductance load (conductance per length)
h	Length, height
$\underline{H}(\omega)$	Transfer function
i	Electric current
I	Sound intensity (sound power per area)
j	Unit of the imaginary numbers ($j^2 = -1$)
$\mathbf{J}_N$	*Bessel* function of the first kind, order N
k	stiffness
k_f	*Fresnel* number
K	Compression module, $K = 1/\kappa$
l	Length
L	Inductance
L'	Inductance load (inductance per length)
L_I	Sound-intensity level
L_p	Sound-pressure level
L_P	Sound-power level
m	Mass
m'_a	Acoustic-mass load (acoustic mass per length)
M	Magnetic-field-transducer coefficient
n	Compliance
n'_a	Acoustic-compliance load (acoustic compliance per length)
N	Electric-field-transducer coefficient

Roman-Letter Symbols ... Continued

p	Sound pressure
P	Power
$\overline{P}$	Active (resistive) power
q	Volume velocity
q_0	Source strength
q_m	Mirror-source strength
q'	Source-strength load (volume-velocity per length),
Q	Sharpness-of-resonance factor (quality factor), also: reactive power
Q_{el}	Electric charge
r	Damping (fluid damping), also: distance, also: radius
$\underline{r}$	Reflectance
r_c	Critical distance
r_{rad}	Radiation resistance
R	Sound-reduction index, also: electric resistance, also: radius
R'	Electric resistance load (resistance per length)
R_I	Isolation index
s	Strain
$\underline{s}$	Complex frequency, $\underline{s} = \breve{\alpha} + j\omega$
S	Standing-wave ratio, also: area of a wall, window, etc.
T	Reverberation time, also: temperature, also: torch, also: period duration
T'	Torch per width
T_{ip}	Transfer coefficient of a transducer, driven as sender
T_{pu}	Transfer coefficient of a transducer, driven as receiver (sensitivity)
T_{pp}	Sound-pressure-transfer factor
T_{up}	Transfer coefficient of a transducer, driven as a sender
u	Electric voltage
U	Perimeter
v	Particle velocity
V	Volume
W	Energy, or work
W'	Energy load (energy per length)
W''	Energy density (energy per volume)
x_{ff}	Far-field distance
Y	Admittance, $Y = 1/Z$
Y_a	Acoustic admittance
Y_f	Field admittance
Y_{mech}	Mechanical admittance
Z	Impedance
Z_0	Terminating acoustic impedance of a tube
Z_a	Acoustic impedance
Z_f	Field impedance
Z_L	Line impedance (acoustic impedance of a tube)
Z_{mech}	Mechanical impedance
Z_{rad}	Radiation impedance
Z_w	Characteristic field impedance of a medium (wave impedance)
Z_{wall}	Wall impedance

Greek-Letter Symbols

α Degree of absorption

$\bar{\alpha}$ Degree of absorption for diffuse sound incidence

$\breve{\alpha}$ Attenuation coefficient

β Phase coefficient (in physics often called angular wave number, k)

$\underline{\gamma}$ Complex propagation coefficient, $\underline{\gamma} = \alpha + j\,\beta$

$\underline{\Gamma}$ Directional characteristics of a sound source or receiver

δ Elevation (vertical angle in a spherical coordinate system),
also: damping coefficient

ε Dielectric permittivity

ε_0 Permittivity of the vacuum

ϵ Flare coefficient

η Ratio of the specific heat capacities, $c_\mathrm{p}/c_\mathrm{v}$, usually denoted γ

Θ Angle of oblique sound incidence

θ Normal (stretching) stress

ϑ *Kronecker* symbol in $\vartheta(z)$ ... *Dirac* impulse [usually denoted $\delta(x)$]

κ Volume compressibility, $\kappa = \kappa_- + \kappa_\sim \approx \kappa_-$

λ Wave length

μ_0 Permeability of the vacuum

μ_d Dipole momentum

$\varXi$ Flow resistivity

ξ Particle displacement

ρ Distance, also: position

ϱ (Mass-)Density, $\varrho = \varrho_- + \varrho_\sim \approx \varrho_-$

σ Electric polarization, also: porosity

ς Mouth correction of a *Helmholtz* resonator

τ Time interval

τ_ph Phase delay

ν Number of turns of a coil

ν_z Number of *Huygens-Fresnel* zones

ϕ Phase angle of a sinusoidal signal

φ Azimuth, i.e. horizontal angle in a spherical coordinate system

Φ Magnetic flux, also: vector potential

Ψ Logarithmic frequency interval

χ Structure factor for porous media

ω Angular frequency

Ω Spherical angle

Specific Mathematical Notations and Terms

$\hat{z}$	Amplitude, peak value
$\underline{z}$	Complex amplitude – note that the $\wedge$ on top of z is omitted
$\overline{z}$	Time average, e.g., used for active power and active intensity
$\overrightarrow{z}$	Vector
$\mid \overrightarrow{z} \mid$, z	Magnitude of a vector
$\mid \underline{z} \mid$, z	Magnitude of a complex amplitude
$\mid \underline{\overrightarrow{z}} \mid$, z	Magnitude of a complex vector
Re$\{\underline{z}\}$	Real-part operator, Re$\{\underline{z}\}$ = Re$\{a+\mathrm{j}b\}$ = a
z_-	Steady component of a function $z(t) = z_- + z_\sim$
$z_\sim$	Alternating component of a function $z(t) = z_- + z_\sim$
$\underline{z}_\rightarrow$	Forward-propagating wave component
$\underline{z}_\leftarrow$	Backward-propagating wave component
z_{rms}	Root of the time average of $z(t)^2$ (root mean square)

For periodic functions of period duration T we have

$$z_{\mathrm{rms}} = \sqrt{\tfrac{1}{T}\int_0^T z(t)^2 \mathrm{d}t}\,,\ \text{otherwise,}\ z_{\mathrm{rms}} = \sqrt{\lim_{T\to\infty}\tfrac{1}{T}\int_0^T z(t)^2 \mathrm{d}t}$$

○—●	... *Fourier* transform
$\sim$	... Proportional
Coefficient	... Multiplier of dimension $\neq$ one
Degree	... Multiplier of dimension one and values of $0-1$ $(0-100\,\%)$
Factor	... Multiplier of dimension one

Units*)

Basic SI-units

m	*Meter* ... unit of length
kg	*Kilogram* ... unit of mass
s	*Second* ... unit of time
A	*Ampere* ... unit of electric current

Some often used SI-derived units

Hz = 1/s	*Hertz* ... unit of frequency
N = (kg m)/s^2	*Newton* ... unit of force
Pa = N/m^2	*Pascal* ... unit of pressure
W = V A = (N m)/s	*Watt* ... unit of power
V = W/A	*Volt* ... unit of electric potential difference, voltage
C = A s	*Coulomb* ... unit of electric charge
Ω = V/A	*Ohm* ... unit of electric resistance to direct current
F = C/V	*Farad* ... unit of capacitance
H = (V s)/A	*Henry* ... unit of self-inductance
T = N/(A m)	*Tesla* ... unit of magnetic flux

*) All units used in this book are coherent with the SI-system (*système international d'unités*)

Index